Selected Titles in This Series

71 **Lajos Pukánszky,** Characters of connected Lie groups, 1999

70 **Carmen Chicone and Yuri Latushkin,** Evolution semigroups in dynamical systems and differential equaitons, 1999

69 **C. T. C. Wall (A. A. Ranicki, Editor),** Surgery on compact manifolds, Second Edition, 1999

68 **David A. Cox and Sheldon Katz,** Mirror symmetry and algebraic geometry, 1999

67 **A. Borel and N. Wallach,** Continuous cohomology, discrete subgroups, and representations of reductive groups, Second Edition, 1999

66 **Yu. Ilyashenko and Weigu Li,** Nonlocal bifurcations, 1999

65 **Carl Faith,** Rings and things and a fine array of twentieth century associative algebra, 1999

64 **Rene A. Carmona and Boris Rozovskii, Editors,** Stochastic partial differential equations: Six perspectives, 1999

63 **Mark Hovey,** Model categories, 1999

62 **Vladimir I. Bogachev,** Gaussian measures, 1998

61 **W. Norrie Everitt and Lawrence Markus,** Boundary value problems and symplectic algebra for ordinary differential and quasi-differential operators, 1999

60 **Iain Raeburn and Dana P. Williams,** Morita equivalence and continuous-trace C^*-algebras, 1998

59 **Paul Howard and Jean E. Rubin,** Consequences of the axiom of choice, 1998

58 **Pavel I. Etingof, Igor B. Frenkel, and Alexander A. Kirillov, Jr.,** Lectures on representation theory and Knizhnik-Zamolodchikov equations, 1998

57 **Marc Levine,** Mixed motives, 1998

56 **Leonid I. Korogodski and Yan S. Soibelman,** Algebras of functions on quantum groups: Part I, 1998

55 **J. Scott Carter and Masahico Saito,** Knotted surfaces and their diagrams, 1998

54 **Casper Goffman, Togo Nishiura, and Daniel Waterman,** Homeomorphisms in analysis, 1997

53 **Andreas Kriegl and Peter W. Michor,** The convenient setting of global analysis, 1997

52 **V. A. Kozlov, V. G. Maz′ya, and J. Rossmann,** Elliptic boundary value problems in domains with point singularities, 1997

51 **Jan Malý and William P. Ziemer,** Fine regularity of solutions of elliptic partial differential equations, 1997

50 **Jon Aaronson,** An introduction to infinite ergodic theory, 1997

49 **R. E. Showalter,** Monotone operators in Banach space and nonlinear partial differential equations, 1997

48 **Paul-Jean Cahen and Jean-Luc Chabert,** Integer-valued polynomials, 1997

47 **A. D. Elmendorf, I. Kriz, M. A. Mandell, and J. P. May (with an appendix by M. Cole),** Rings, modules, and algebras in stable homotopy theory, 1997

46 **Stephen Lipscomb,** Symmetric inverse semigroups, 1996

45 **George M. Bergman and Adam O. Hausknecht,** Cogroups and co-rings in categories of associative rings, 1996

44 **J. Amorós, M. Burger, K. Corlette, D. Kotschick, and D. Toledo,** Fundamental groups of compact Kähler manifolds, 1996

43 **James E. Humphreys,** Conjugacy classes in semisimple algebraic groups, 1995

42 **Ralph Freese, Jaroslav Ježek, and J. B. Nation,** Free lattices, 1995

41 **Hal L. Smith,** Monotone dynamical systems: an introduction to the theory of competitive and cooperative systems, 1995

(Continued in the back of this publication)

Characters of Connected Lie Groups

Mathematical Surveys and Monographs

Volume 71

Characters of Connected Lie Groups

Lajos Pukánszky

American Mathematical Society

1991 *Mathematics Subject Classification.* Primary 22E45.

ABSTRACT. The central topic of this book, the theory of characters of general connected Lie groups, is treated here for the first time in book form, in a complete and compact form. This is also the first systematic exposition of the orbit method in representation theory in non-type I situations.

Library of Congress Cataloging-in-Publication Data

Pukánszky, L.
Characters of connected Lie groups / Lajos Pukánszky.
p. cm. — (Mathematical surveys and monographs, ISSN 0076-5376 ; v. 71)
Includes bibliographical references.
ISBN 0-8218-1088-X (acid-free paper)
1. Lie groups. 2. Characters of groups. 3. Representations of groups. I. Title. II. Series: Mathematical surveys and monographs ; no. 71.
QA387.P84 1999
512′.55—dc21 99-29967
CIP

∞ The paper used in this book is acid-free and falls within the guidelines
established to ensure permanence and durability.
Visit the AMS home page at URL: http://www.ams.org/

10 9 8 7 6 5 4 3 2 1 04 03 02 01 00 99

Contents

Preface

Lajos Pukánszky died in February 1996. He was preparing a book, "Characters of Connected Lie Groups". His plan, consisting of six chapters, was found in his papers:

I. Unitary representations of locally algebraic groups

We start by showing that they are of type I. This result is going to be used throughout the book. This would be followed by a result, referred to as the Main Lemma, on the action of an algebraic group on the dual of another. The chapter would close with an application of the Main Lemma to prove Dixmier's theorem, that the regular representation of any connected Lie group is semifinite.

II. Unitary representations of elementary groups

An elementary group is a central extension by a one-dimensional torus, of a closed subgroup of $\mathbb{R}^n$. The content of this chapter is auxiliary material, to be used in the subsequent chapters.

III. Existence of characters

This is the core of the book. The principal result asserts, that by assigning to a quasi-equivalence class of normal representation its kernel, one obtains a bijection with the underlying set of the primitive ideal space of the connected Lie group under consideration.

IV. Generalized Kirillov theory

Here our principal objective is to extend Kirillov's theory of the unitary representations of nilpotent groups to any connected and simply connected solvable Lie group. The result is a canonical bijection between the underlying set of the primitive ideal space, and members of a family of geometric objects called generalized orbits. In the simplest case, as that of nilpotent groups, these reduce to the collection of all coadjoint orbits.

V. The character formula

Here our objective is an extension of the character formula of Kirillov, to any connected solvable Lie group.

VI. Generalized orbits and symplectic geometry

This is dedicated to an investigation per se of the generalized orbits. Question: What features, given a differentiable manifold, will turn it into a generalized orbit?

Pukánszky could not write chapters V and VI. On the other hand, he wrote an almost final version of chapters I–IV. Since these chapters form a coherent set, their publication was absolutely necessary. We hope that the present book realizes, at least partially, what Pukánszky had wanted.

J. Dixmier
M. Duflo

Foreword

This book is meant to be a contribution of the line of research which, going back to A. Weil, was taken up again by R. Godement, and continued by G. W. Mackey, J. Dixmier, A. A. Kirillov and many others. The common theme of all this is unitary representations of locally compact groups, often viewed from very different angles. In its approach it produces an interplay between classical group theory and modern analysis. More specifically, in this volume we shall deal with unitary representations of connected Lie groups, which are not semi-simple. Among others, we shall prove an analogue of the classical bijection between prime ideals of a finite group and characters of irreducible unitary representations (cf. Chapter III). Chapter IV confines itself to connected solvable Lie groups and provides a generalization of Kirillov's theory of unitary representations of nilpotent Lie groups.

To close this foreword it must be frankly admitted that the reading of this book requires much preliminary background. It is assumed that the reader is familiar with the introductory chapter to [**AM**], and the relevant paragraphs of the last section of [**D***] (letters in brackets refer to the bibliography at the end).

This volume is dedicated to R. V. Kadison, in honor of his 70th birthday and in commemoration of many years as colleagues at the University of Pennsylvania.

Introduction

The problems treated in this book can be illustrated by simple examples.

A) Let us consider first the universal covering G of the motion group of $\mathbb{R}^2$. It can be realized as the set of all pairs (t, z) $(t \in \mathbb{R}, z \in \mathbb{C})$ with the law of multiplication

$$(t, z)(t', z') = (t + t', z + e^{it} z').$$

We shall next be concerned with the definition and determination of the characters of G. Since G is solvable, by virtue of the theorem of Lie, any irreducible unitary representation of a dimension greater than one is infinite-dimensional. It can be shown that, up to unitary equivalence, all these can be obtained by the following procedure. Let us consider the abelian subgroup

$$W = \{(2\pi n, z);\ n \in \mathbb{N}, z \in \mathbb{C}\}.$$

Let α be a fixed real number and z_0 a nonzero complex number. We can define a character χ of W by

$$\chi((2\pi n, z)) = \exp(i\alpha n)\exp(i \operatorname{Re}(z\bar{z}_0)).$$

To any such character we can associate a unitary representation T of G as follows: We shall write, whenever convenient, t in place of $(t, 0)$ and z in place of $(0, z)$. Let us write $I = \{-\pi \leqq t \leqq \pi\} \subset G$. We have setwise $G = WI$. Let us take, as the space $H(T)$ of T, $L^2(I)$ with the Lebesgue measure. For $f \in H(T)$ and $a \in G$ we put

$$(T(a)f)(t) \equiv \chi(z(ta))f(t\bar{a}) \qquad (t \in I)$$

where $z(ta) \in W$ and $t\bar{a} \in I$ are such that $z(ta) \cdot t\bar{a} = ta$. T is the unitary representation of G induced by the character $\chi \in \mathrm{X}(W)$. The above form is convenient for the computation of the character of T. As such, we propose to take the functional $\varphi \mapsto \operatorname{Tr}(T(\varphi))$, where for $\varphi \in C_c^\infty(G)$

$$T(\varphi) = \int_G \varphi(a)T(a) \cdot da$$

with $da = dt\,dz$, which is just the bi-invariant Haar measure. Taking a positive definite φ, we show that this is meaningful, as follows: $T(\varphi)$ can easily be shown to be implemented on $L^2(I)$ by the kernel K_φ defined as

$$K_\varphi(\tau, t) = \int_W \varphi(\tau^{-1}wt)\,dw.$$

If the support of φ is small enough then, writing Φ for the properly normalized Fourier transform of $\varphi|W_0$, and setting $T_t z_0 = e^{it}z_0$, we obtain

$$K_\varphi(\tau, \tau) = \Phi(T_\tau z_0)$$

and thus

$$\mathrm{Tr}(T(\varphi)) = \int_I K_\varphi(\tau,\tau)\,d\tau = \int_I \Phi(T_\tau z_0)\,d\tau.$$

Obviously, we have no problem with the convergence of the expression on the right.

B) Let us assume next that λ is a fixed irrational number and define a law of multiplication on the set (t, z_1, z_2) $(t \in \mathbb{R}; z_1, z_2 \in \mathbb{C}\}$ by

$$(t, z_1, z_2)(t', z_1', z_2') = (t + t', z_1 + e^{it} z_1', z_2 + e^{i\lambda t} z_2').$$

Let us write M for the group so obtained. It is known to be the connected Lie group of lowest dimension, which is not of type I. Next we shall try to imitate for M the computation of characters, introduced in A) for G.

We write $I = \{(t,0,0); t \in \mathbb{R}\}$ and $W = \{(0, z_1, z_2); z_1, z_2 \in \mathbb{C}\}$. Then we have setwise $G = WI$. We take a pair of nonzero complex numbers u_1, u_2 and define for $w = (w_1, w_2)$

$$\chi(w) = \exp(i\,\mathrm{Re}(w_1\overline{u}_1 + w_2\overline{u}_2)).$$

Next we form the unitary representation T, induced by χ, of M. This admits a description similar to what we saw in A). For $f \in H(T) = L^2(\mathbb{R})$ and $a \in G$ we put

$$(T(a)f)(t) \equiv \chi(w(ta))f(t\overline{a})$$

where $w(ta) \in W$ and $t\overline{a} \in I$ are such that $ta = w(ta)\cdot t\overline{a}$. With notation analogous to those of A) we form, for $\varphi \in C_c^\infty(G)$, the operator $T(\varphi)$. It is implemented on $L^2(I)$ by the kernel

$$K_\varphi(\tau, t) = \int_W \varphi(\tau^{-1} w t)\chi(w)\,dw.$$

Writing $T_t(w_1, w_2) = (e^{it}w_1, e^{i\lambda t}w_2)$ and Φ for the Fourier transform of $\varphi|W$, we obtain $(u = (u_1, u_2))$

$$K_\varphi(t,t) = \Phi(T_t u).$$

To obtain $\mathrm{Tr}(T(\varphi))$, we would have to take the integral of the expression on the right over the whole real line. But, of course, in general, it fails to converge. Instead of this, one is tempted to consider

$$\lim_{T\to+\infty} \frac{1}{2T}\int_{-T}^{T} \Phi(T_t u)\,dt$$

which is indeed, as far as convergence is concerned, a viable possibility, but it does not provide the trace of $T(\varphi)$. A more thorough analysis reveals that there is a factor representation U of M such that the above expression provides the trace of $U(\varphi)$ in the sense of, in general, not of type I factor, generated by U. Let $T(u)$ be the representation of M considered above. Then U can be shown to be

$$U = \int_O^\oplus T(u)\cdot du$$

where O is the closure of $\{T_\tau u; \tau \in \mathbb{R}\}$ and du is an absolutely continuous measure on O.

This being so, the content of this book can be described as follows. Let G be a connected Lie group. Following R. Godement, a factor representation T of G

will be called normal if it generates a semifinite factor such that there is a nonzero element φ in $C^*(G)$ and such that $T(\varphi)$ admits a trace.

By virtue of a result of J. Y. Charbonnel, this is the case if and only if there is $\varphi \in C_c^\infty(G)$ such that $T(\varphi)$ is traceable. Let A be a separable C^*-algebra and $\mathrm{Prim}(A)$ the set of closed 2-sided primitive ideals of A. As J. Dixmier has shown, for any factor representation of A, its kernel belongs to $\mathrm{Prim}(A)$. If $A = C^*(G)$ then we write for this $\mathrm{Prim}(G)$.

It is inviting to consider $C^*(G)$ as the analogue, for any locally compact group, of the algebra of a finite group. By taking the hint provided by the finite groups, we can seek a bijection between $\mathrm{Prim}(G)$ and the set $\widehat{G}_{\mathrm{norm}}$ of the quasi-equivalence classes of normal representations of G. That this is indeed so for connected Lie groups is the main result of his book (cf. Ch. III). Let us note, incidentally, that the analogous statement, for any locally compact group, fails to be true. For the other main objectives, see the introduction to Chapter IV.

Some notation used throughout the book

If A is an algebra, or a group, or a Lie algebra, A^{c} denotes the center of A.

If H is a locally compact group, H_0 denotes the connected component of the unit element. $\widetilde{H}$ denotes some "algebraic envelope" of H (see later). $C^*(H)$, or sometimes $\mathfrak{H}$, denotes the C^*-algebra of H. $\mathrm{X}(H)$ denotes the group of unitary continuous characters (one-dimensional) of H. $\mathrm{Prim}(H)$ denotes the set of primitive ideals of $C^*(H)$.

If S is a locally compact space, $\mathcal{L}(S)$ denote the set of complex continuous functions with compact support on S.

$\mathbb{N}, \mathbb{Z}, \mathbb{Q}, \mathbb{R}, \mathbb{C}, \mathbb{T}$ have the usual meaning.

CHAPTER I

Unitary Representations of Locally Algebraic Groups

1.1. On a theorem of Chevalley

In this work, an important role will be played by a proposition due to C. Chevalley. To state it we recall that a Borel space is called countably separated, if it contains a sequence of Borel sets such that each point is the intersection of all those members of our sequence in which it is contained. This being so, let G be a real linear algebraic group acting on the finite-dimensional real vector space V by a representation.

LEMMA 1. *With the above notation, let G' be an open subgroup of G. Then V/G' is countably separated.*

This statement is a special case of the following proposition. For its proof the reader is referred to [**Dh**, p. 183].

LEMMA 2. *Let V be an algebraic variety, G a real algebraic group acting regularly on V. If G' is an open subgroup of G, then V/G' is countably separated.*

It is important to tie the property of being countably separated to other properties of the orbit space. An important result in this direction is accomplished by a theorem of J. Glimm which we now recall (cf. [**Gl**], Theorem 1, (1) (2) (3), p. 124). Let B be a topological space. A subset S of B is called quasi-compact if every open cover of S contains a finite subcover. B itself is said to be locally quasi-compact if any of its points admits a fundamental system of quasi-compact neighborhoods. We say that a locally quasi-compact space B is almost Hausdorff if there is an ordinal γ and a family $(O_\alpha)_{0\le\alpha\le\gamma}$ of ascending open subsets of B, such that $O_0=\varnothing$ and $O_\gamma=B$; if δ is a limit ordinal $\le\gamma$, then O_δ is a union of $(O_\beta)_{\beta<\delta}$; if $\delta=\alpha+1$, then $O_{\alpha+1}-O_\alpha$ is a dense Hausdorff subset of $B-O_\alpha$.

Important examples of spaces that are almost Hausdorff are provided by the duals of the type I C^*-algebras.

This being so, assume that B is separable, almost Hausdorff, and is acted upon by a separable, locally compact group G as a transformation group. Then the following holds true.

LEMMA 3. *With the above definitions and assumptions, B/G is countably separated if and only if for each $m\in B$, the orbit Gm is open in its closure.*

1.2. Locally algebraic groups

We recall that a separable locally compact group G is said to be of type one if any of its factor representations generates a von Neumann algebra that is *-isomorphic to a finite or infinite-dimensional full ring. For such a representation we shall write $T = (I)$, and similarly $G = (I)$ will indicate that G is a type I group. Connected Lie groups fail to be of type I already in five dimensions. The first major result, characterizing important classes of connected Lie groups that are of type I, is due to Harish-Chandra (cf. [**H**]). He showed that any connected semi-simple Lie group is of type I. Following this, Dixmier proved that any real algebraic group is of type I (cf. [**Da**]). Let us note that his proof assumes the theorem of Harish-Chandra. Dixmier also observed that there are connected solvable Lie groups, which are of type I, while their universal covering fails to have this property. It was remarked, however, that the said implication is valid if our group is the connected component of the identity of a solvable algebraic group. Subsequently, Dixmier proved that here the adjective "solvable" can be dropped (cf. [**Dreg**], Proposition p. 425, in particular). The objective of this section is to present a demonstration of this proposition. It assumes Harish-Chandra's result, but not Dixmier's original theorem. We start with the following

DEFINITION. Let G be a connected Lie group. It will be called locally algebraic if it is locally isomorphic to a real algebraic group. In this case we shall write $G = (a)$.

This being so, we have the following

THEOREM. *Let G be a connected Lie group that is locally algebraic. Then G is of type one.*

The proof below is much influenced by that of Dixmier. It proceeds by induction, according to the dimension of the group. Evidently, we can assume that G is simply connected.

(i) DEFINITION. Let V be a real vector space of finite dimension. Let $\mathfrak{g}$ be a Lie algebra of endomorphisms acting on V. Then $\mathfrak{g}$ is said to be algebraic if it belongs to an algebraic subgroup of $\mathrm{GL}(V)$. In this case we shall write $\mathfrak{g} = (a)$.

We recall (cf. [**Ch III**], Proposition 5, p. 144) that a Lie algebra of endomorphisms is algebraic if and only if there are subalgebras $\mathfrak{i}, \mathfrak{c}, \mathfrak{m}$ such that $\mathfrak{g} = \mathfrak{i} + \mathfrak{c} + \mathfrak{m}$ (direct sum of subspaces); here $\mathfrak{i}$ is semi-simple, $\mathfrak{c}$ is an abelian Lie algebra of semi-simple endomorphisms, such that $[\mathfrak{i}, \mathfrak{c}] = 0$ and $\mathfrak{c}$ is algebraic, which means that in the complexification of V, there is a basis of eigenvectors with integral eigenvalues; finally, $\mathfrak{m}$ is the largest ideal, composed solely of nilpotent endomorphisms; it is well-determined. Below we can assume that $\mathfrak{m}$ is nonzero. In fact, if $\mathfrak{m} = \{0\}$, then $\mathfrak{g}$ is reductive and the desired conclusion is trivially implied by the theorem of Harish-Chandra quoted above.

(ii) From now on we shall assume that $\dim(\mathfrak{m}) > 0$. Note that in this case there is a nontrivial abelian $\mathfrak{g}$-ideal in $\mathfrak{m}$; e.g., one can take the center of $\mathfrak{m}$. Below we shall distinguish two cases: A) $\mathfrak{m}$ contains an abelian $\mathfrak{g}$-ideal $\mathfrak{a}$, such that either $\dim(\mathfrak{a}) > 1$ or $\dim(\mathfrak{a}) = 1$, but $\mathfrak{a}$ is not contained in the center of $\mathfrak{g}$; B) the condition of A) cannot be fulfilled.

A) Assume that G_1 is the irreducible (in the sense of algebraic geometry) subgroup of $\mathrm{GL}(V)$, giving rise to the Lie algebra $\mathfrak{g}$. We can define for $a \in G_1$ the endomorphism $\rho(a)$ of $\mathfrak{a}$ by

$$\rho(a)l = ala^{-1}.$$

Let us write $\overline{G}$ for the connected component of the unity in G_1. Recalling that G is simply connected, we denote by τ the canonical morphism from G onto $\overline{G}$.

We note that $\rho(\overline{G})$ is the connected component of the identity of some algebraic group in $\mathrm{GL}(\mathfrak{a})$. In fact, let us write σ for the differential of ρ. We have to show that $\sigma(\mathfrak{g}) = (a)$. We recall that by what we said in (i), we can write $\mathfrak{g}$ as $\mathfrak{i}+\mathfrak{c}+\mathfrak{m}$, with properties described there. Then clearly $\sigma(\mathfrak{i}), \sigma(\mathfrak{c})$, and $\sigma(\mathfrak{m})$ are all algebraic in $\mathfrak{gl}(\mathfrak{a})$. Next we recall the following result (cf. [**Ch II**], Théorème, p. 175). Given a family of algebraic Lie algebras on a vector space, the Lie algebra they generate is again algebraic. From this we conclude that $\sigma(\mathfrak{g})$ is algebraic. Let G' be an algebraic group in $\mathrm{GL}(\mathfrak{a})$, giving rise to $\sigma(\mathfrak{g})$. Then $\rho(\overline{G}) = (G')_0$ and thus $\rho(\overline{G})$ is the connected component of the unity of an algebraic group in $\mathrm{GL}(\mathfrak{a})$, as claimed above. Similar conclusions hold true for the representation, contragredient to ρ, in $\mathfrak{a}^*$. We write A for the connected and simply connected group, corresponding to $\mathfrak{a}$, in G. We shall consider A^* as a G-space, the G-action here being contragredient to the G-action on A considered above. We recall that by what we saw in 1.1, A^*/G is countably separated. Let λ be a factor representation of G; we want to show that it is of type I. The restriction of λ to A gives rise to an ergodic measure on A^*, and thus this measure is carried by a G-orbit Ω (say). Below we shall distinguish two subcases: A1) $\dim(\Omega) > 0$; A2) $\dim(\Omega) = 0$.

(iii) A1)

a) Let ω be an element in Ω; then we have $\Omega = G\omega \subset A^*$. We note that $(G_\omega)_0$ is locally algebraic. In fact, to see this, it is enough to recall that G acts on A^* as does the connected component of the unity of an algebraic group. From this we conclude that since $\dim(\Omega) > 0$, we have $\dim((G_\omega)_0) < \dim(G)$ and thus by induction we infer that $(G_\omega)_0$ is of type I.

b) Next we invoke Mackey's Imprimitivity Theorem (cf. [**AM**], Theorem 6, p. 58). In the present case it implies the following. We have $\lambda = \mathrm{ind}_{G_\omega \uparrow G}\, \mu$, where μ is a factor representation of G_ω, the type of which equals that of λ. Writing $S = G_\omega$, it is thus sufficient to show that S is of type I. We recall (cf. a)) that S_0 is of type I.

c) Let us put $S_1 = \tau^{-1}((\overline{G}_\omega)_0)$. We have $\tau(S_1) = \tau(S_0) = (\overline{G}_\omega)_0$, from where we conclude that there is a central subgroup L of G such that $S_1 = LS_0$. In this manner S_1, too, is of type I.

d) We deduce from all this that S/S_1 is finite. In fact, it is isomorphic to $\overline{G}_\omega/(\overline{G}_\omega)_0$. (We recall that $\overline{G} = (G_1)_0 \subset \mathrm{GL}(V)$.) Hence it suffices to establish the following

PROPOSITION. *Let G be a separable, locally compact group. Let H be a closed, invariant subgroup of finite index of G, which is of type I. Then G, too, is of type I.*

In fact, our assumptions imply that $\widehat{H}$ is a standard Borel space on which G acts as $G/H = F$, say; hence F is finite. Evidently, the actions of elements of F are Borel automorphisms of $\widehat{H}$. We note that $\widehat{H}/F$ is countably separated. In fact, this is immediately implied by the following statement due to Varadarajan (cf. [**AM**], Proposition 3.4, p. 15). Assume that E is an analytic Borel G-space, where G is a separable locally compact group. Then there is a Polish space $\mathcal{F}$, on which G acts as a topological transformation group; there is a G-invariant Borel subset B of $\mathcal{F}$, and a G-equivariant Borel isomorphism between E and B. Suppose that ζ is a factor representation of G. To show that ζ is of type I, we proceed very much as in b) above. Since $\widehat{H}/G$ is countably separated, $\zeta|H$ is carried by a G/H-orbit O, say. Assume that $x \in O$ is fixed. Then ζ can be obtained through the following procedure. There is a projective extension y of x to G_x, such that if $\iota : G_x \to G_x/H$ is the canonical morphism, the obstruction cocycle can be assumed to be of the form $(\iota(\alpha))^{-1}$. There is an α-factor representation z of G_x/H such that

$$\zeta = \mathrm{ind}_{G_x \uparrow G}(y \otimes z)^{\sim}.$$

The type of ζ is the same as that of z. But z identifies to a factor representation of a central extension of a finite group by a 1-torus. The whole being compact, z is of type I. Summing up, we have completed proving that if G is a separable, locally compact group containing a closed invariant type I subgroup H such that G/H is finite, then G itself is of type I.

e) We have completed proving the following. If $\omega \in \widehat{A}$ is such that $\dim(G_\omega) < \dim(G)$, then if λ is a factor representation of G such that its restriction to A decomposes on $G\omega = \Omega$, $\dim(\Omega) > 0$, then λ is of type I.

(iv) A2)

Here we assume that $\dim(\Omega) = 0$; in this case $\Omega = \{\omega\}$. Let us start by noting that if $J = (\ker(\omega))_0$, then the dimension of J is positive. This is obvious if the dimension of $\mathfrak{a}$ exceeds one. Assuming that the dimension of $\mathfrak{a}$ is equal to one, we choose a nonzero element j in $\mathfrak{a}$. We have then $aj \equiv \mu(a)j$ $(a \in G)$ where, since j does not belong to the center, μ is not identically one. We have

$$\begin{aligned}\omega(\exp(tj)) &= (a\omega)(\exp(tj)) = \omega(a^{-1}\exp(tj)a)\\ &= \omega(\exp((t/\mu(a)))j) \quad \text{for all } t \in \mathbb{R}\end{aligned}$$

whence $\omega = 1$, and $J = A$. Let us put $G' = G/J$ and write ϕ for the canonical morphism $G \to G'$. There is a factor representation λ' of G' such that $\lambda = \lambda' \circ \phi$. We write $\mathfrak{g}' = \mathrm{Lie}(G')$ and observe that by virtue of the assumption of our inductive procedure, it is enough to show that $\mathfrak{g}'$ is algebraic.

We continue by recalling a few things concerning algebraic groups.

1) Let $\mathfrak{g} = (a)$, and ρ a linear representation of $\mathfrak{g}$ on a finite-dimensional real vector space. We say ([**Ch III**], p. 47) that ρ is a rational representation of $\mathfrak{g}$ if it is the differential of a rational representation of an algebraic group belonging to $\mathfrak{g}$.

2) (cf. [**Ch III**], Corollary 1, p. 48). Assume that $\mathfrak{g}$ is an algebraic Lie algebra and that $\mathfrak{h}$ is of the same kind, an ideal in $\mathfrak{g}$. If ρ is a rational representation of $\mathfrak{g}$, then $\rho(\mathfrak{h})$ is algebraic. Hence, in particular, $\rho(\mathfrak{g})$ is algebraic.

3) (cf. [**Ch III**], p. 119). Assume that G and H are algebraic groups such that H is invariant in G. Then there is a rational representation of G, the kernel of which is equal to H.

We conclude from all this that there is a rational representation σ of $\mathfrak{g}$ such that $\ker(\sigma) = \mathfrak{a}$ and thus it is enough to take $\mathfrak{g}' = \sigma(\mathfrak{g})$.

(v) To prove that if G is connected and simply connected and locally algebraic, then G is of type I, it remains to consider the case when the condition of A) cannot be realized. Then we have $\dim(\mathfrak{m}^c) = 1$, and $\mathfrak{m}^c \subset \mathfrak{g}^c$. We can assume that $\lambda|\mathfrak{m}^c$ is not identically one.

In fact, otherwise we can reason as in (iv) above.

a) We show first that if the dimension of $\mathfrak{m}$ is equal to one, then G is of type one.

In fact, in this case, since $\mathfrak{m}$ is in the center, we conclude that $\mathfrak{g}$ is reductive, and thus we can complete our discussion as in (i) above. Therefore from now on we can assume that $\dim(\mathfrak{m}) > 1$.

b) We say that the Lie algebra $\mathfrak{g}$ is Heisenberg and write $\mathfrak{g} = (H)$ if $[\mathfrak{g}, \mathfrak{g}] = \mathfrak{g}^c$ and $\dim(\mathfrak{g}^c) = 1$. We claim that we have $\mathfrak{m} = (H)$. In fact, since now $\dim(\mathfrak{m}^c) = 1$, it is enough to show that $[\mathfrak{m}, \mathfrak{m}] = \mathfrak{m}^c$. In fact, if not, then we have $\dim([\mathfrak{m}, \mathfrak{m}]) \geq 2$, and there are ideals $\mathfrak{m}_k$ $(k = 1, 2)$ in $\mathfrak{m}$ such that $\dim(\mathfrak{m}_k) = k$ and

$$[\mathfrak{m}, \mathfrak{m}] \supseteq \mathfrak{m}_2 \supset \mathfrak{m}_1.$$

We have

$$[[\mathfrak{m}, \mathfrak{m}], \mathfrak{m}_2] \subseteqq [[\mathfrak{m}_2, \mathfrak{m}], \mathfrak{m}] = 0.$$

Hence $\mathfrak{m}_2 \subseteqq [\mathfrak{m}, \mathfrak{m}]^c$. But $([\mathfrak{m}, \mathfrak{m}])^c$ is an abelian $\mathfrak{g}$-ideal in $\mathfrak{m}$, and thus we are in contradiction with the assumption that A) cannot be realized. Thus we can assume that $\mathfrak{m} = (H)$.

c) Let us write H for the connected and simply connected subgroup corresponding to $\mathfrak{m}$. Then by $\mathfrak{m} = (H)$, there is an irreducible representation ω of H such that $\dim(\omega) > 1$. We recall that, up to unitary equivalence, ω is well-determined by $\omega|M^c$ (cf. [**N**]). Given the factor representation λ of G, we want to show that $\lambda = (I)$. Let η be the character of M^c to which $\lambda|M^c$ is quasi-equivalent. We can assume $\eta \not\equiv 1$. Hence evidently, since $\mathfrak{m}^c \subseteqq \mathfrak{g}^c$, for any $c \in G$, $c\omega$ is unitarily equivalent to ω, and thus there is a projective extension ω' of ω to G permitting an analysis of λ by aid of some of the projective representations of G/M (cf. [**AM**], p. 61).

d) Because of later applications in §1.3 below, we interrupt now for an explicit construction of the cocycle. The reader, interested only in proving that $\lambda = (I)$, can proceed immediately to read f).

d1) Let h be an element of the dual of $\mathfrak{m}^c$ such that for all l in $\mathfrak{m}^c$ we have

$$\eta(\exp(l)) = \exp[i(l, h)].$$

We conclude from this that $G = G_h M$. Our first objective will be to establish the following. Let $\mathfrak{h}$ be the space of ω. Then there is a unitary representation T of G_h on $\mathfrak{h}$ such that for all m in M and a in G_h we have

$$T(a)\omega(m)T(a^{-1}) = \omega(ama^{-1}).$$

d2) Let c be a fixed nonzero element in $\mathfrak{m}^c$.

Since $[\mathfrak{m}, \mathfrak{m}] = \mathfrak{m}^c$, there is a skew symmetric, bilinear form B_1 on $\mathfrak{m} \times \mathfrak{m}$ such that

$$[x, y] = B_1(x, y)c \quad (x, y \in \mathfrak{m}).$$

By $\mathfrak{m}^c \subseteqq \mathfrak{g}^c$, we have for all $a \in G$

$$B_1(ax, ay) \equiv B_1(x, y) \quad (x, y \in \mathfrak{m}).$$

Let $\mathfrak{m}'$ be the totality of elements in $\mathfrak{m}$, orthogonal to h. Evidently, $\mathfrak{m}'$ is invariant under G_h. Since h is not orthogonal to $\mathfrak{m}^c$, the restriction of B_1 to $\mathfrak{m}' \times \mathfrak{m}'$ $(= B$, say) is nondegenerate.

Let $\mathfrak{r}$ be the subgroup, leaving B invariant, of $\mathrm{GL}(\mathfrak{m}')$. We put $\psi(a) \equiv \mathrm{Ad}(a)|\mathfrak{m}$ $(a \in G_h)$. Then $a \mapsto \psi(a)$ is a representation of G_h in $\mathfrak{r}$.

d3) Let $(\mathcal{R}, p)$ be the universal covering of $\mathfrak{r}$. We claim that there is a morphism ψ' of G_h in $\mathcal{R}$ such that

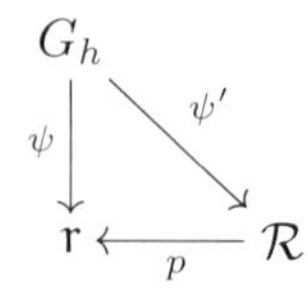

In fact, we recall the following (cf. [**Ch III**], p. 215). Let G be a simply connected and G' a connected Lie group. Let $\widetilde{G}'$ be the universal covering of G' and let $p : \widetilde{G}' \to G'$ be the canonical projection. Then, for each homomorphism $f : G \to G'$, there exists a morphism $g : G \to \widetilde{G}'$, well-determined by $f = p \circ g$.

We conclude from this that it is enough to establish that G_h is simply connected. To see this, we claim that G_h/M_h is connected and simply connected.

Since $G = G_h M$, we have that

$$G/M = G_h M/M = G_h/M_h.$$

Hence, G_h is connected and G_h/M_h is simply connected and thus, since M_h is isomorphic to $\mathbb{R}$, G_h is simply connected.

d4) Let c be a nonzero fixed element of $\mathfrak{m}^{\mathrm{c}}$. We have $\mathfrak{m} = \mathfrak{m}' + \mathbb{R}c$. Let s be a fixed element of $\mathfrak{r}$ $(\subset \mathrm{GL}(\mathfrak{m}'))$. We note that the map $l + tc \mapsto sl + tc$ $(l \in \mathfrak{m}', t \in \mathbb{R})$ defines an automorphism of $\mathfrak{m}$. Writing $\sigma(x + tc) = sx + tc$, we have

$$\begin{aligned}[\sigma(x+tc), \sigma(y+t'c)] &= [sx + tc, sy + t'c] \\ &= B(sx, sy)c = B(x, y)c = \sigma(B(x, y)c) = \sigma([x + tc, y + t'c])\end{aligned}$$

and thus $[\sigma(x + tc), \sigma(y + t'c)] = \sigma([x + tc, y + t'c])$.

Given $s \in \mathfrak{r}$ fixed, we write $m \mapsto m^s$ $(m \in M)$ for the corresponding automorphism of M. We observe that $m^{\psi(a)} = ama^{-1}$ $(a \in G_h, m \in M)$. In fact, assume that $m = \exp(l + tc)$ $(l \in m', t \in \mathbb{R})$. We have

$$ama^{-1} = \exp(al + tc) = \exp(\psi(a)l + tc) = m^{\psi(a)}$$

and thus $ama^{-1} = m^{\psi(a)}$ $(a \in G_h, m \in M)$.

d5) We recall (cf. [**Kir II**] Lemma 3, p. 39) that there is a representation w of $\mathcal{R}$ on $\mathfrak{h}$ ($=$ space of ω) such that

$$w(b)\omega(m)w(b^{-1}) = \omega(m^{p(b)}) \quad (m \in M, b \in \mathcal{R}).$$

We recall that an objective is to find a unitary representation T of G_h on $\mathfrak{h}$ such that for all $a \in G_h$ and $m \in M$ we have

$$T(a)\omega(m)T(a^{-1}) = \omega(ama^{-1}).$$

Conclusion: By what precedes it is enough to set $T(a) = w(\psi'(a))$ $(a \in G_h)$. In fact, we have

$$\begin{aligned}T(a)\omega(m)T(a^{-1}) &= w(\psi'(a))\omega(m)w(\psi'(a^{-1})) \\ &= \omega(m^{p(\psi(a))}) = \omega(m^{\psi(a)}) = \omega(ama^{-1})\end{aligned}$$

whence $T(a)\omega(m)T(a^{-1}) = \omega(ama^{-1})$ $(a \in G_h, m \in M)$.

e) There is a map (in fact a Levi section) $\sigma : G/M \to G$, which is smooth, such that $\pi(\sigma(s)) = s$ $(s \in G/M)$. We observe that we can assume $\sigma(G/M) \subset G_h$. In fact, given $g \in Mh$, there is a unique $f(g) \in \mathfrak{m}'$, satisfying $\exp(f(g))h = g$. To see this we note that evidently there is an l in $\mathfrak{m}$ such that $\exp(l)h = g$. We have also

$\exp(l + ct)h = g$. Hence to achieve $[l, h] = 0$ it is enough to replace, if needed, l by $l + ct$ with an appropriate t. Uniqueness is evident. Summing up, we have

$$\sigma(a)h = \exp(f(\sigma(a)))h = \mu(a)h, \text{say.}$$

Since $\sigma(a)h = \mu(a)h$, it is enough to replace σ by $\mu^{-1}\sigma$. We write $S = \sigma(G/M) \subset G_h$. Let us note that the restriction of π to S is injective. In fact, let us assume that $\pi(s_1) = \pi(s_2)$ $(s_k = \sigma(f_k), f_k \in G/M$ for $k = 1, 2)$. This clearly implies $f_1 = f_2$.

We conclude from this that we can define $s : G \to S$, smooth, such that with $m(a)$ in M we have $a = s(a)m(a)$. We note also that $G = SM$ holds setwise. We observe next that, if $a, b \in G$, then $m(s(a)s(b))$ is in M^c. In fact, it suffices to remark that if $a, b \in G_h$, then $m(ab) \in G_h \cap M = M^c$. We conclude from here that we can define for $a, b \in G : \beta(a, b) = \overline{\eta(m(s(a) \cdot s(b)))}$. We recall that there is a continuous representation T of G_h on $\mathfrak{h}$, such that we have for $a, b \in G_h$ and $m \in M$

$$T(a)\omega(m)T(a^{-1}) = \omega(ama^{-1}).$$

We define $\omega'(a) = T(s(a))\omega(m(a))$. This being so, let us observe that if $a, b \in G$, then

$$\omega'(a)\omega'(b) = \beta(a, b)\omega'(ab).$$

In fact, let us assume that $a = rm, b = tn$ $(r, t \in S; m, n \in M)$. We have

$$\begin{aligned}\omega'(a)\omega'(b) &= T(r)\omega(m)T(t)\omega(n)\\ &= T(r)T(t)(T(t^{-1})\omega(m)T(t))\omega(n)\\ &= T(rt)\omega(t^{-1}mt)\omega(n).\end{aligned}$$

We note that by $T(r) = w(\psi'(r)) : T(r) = 1$ if $r \in M^c$ $(\subset G^c)$. We conclude from this that

$$\omega'(a)\omega'(b) = T(s(rt)) \cdot \omega(t^{-1}mt \cdot n).$$

Let us note that $ab = rm \cdot tn = rt(t^{-1}mt)n = s(rt)m(rt) \cdot t^{-1}mt \cdot n$ which implies that

$$\omega'(ab) = T(s(rt)) \cdot \omega(m(rt)) \cdot \omega(t^{-1}mt \cdot n).$$

Since $m(rt) \in M^c$ and $\omega|M^c \approx \eta$, we obtain

$$\omega'(ab) = \eta(m(rt)) \cdot T(s(rt)) \cdot \omega(t^{-1}mt \cdot n).$$

We recall that $\omega'(a)\omega'(b) = T(s(rt)) \cdot \omega(t^{-1}mt \cdot n)$ from where we conclude that

$$\omega'(a)\omega'(b) = \overline{\eta(m(rt))}\omega'(ab).$$

But $r = s(a), t = s(b)$, whence

$$\omega'(a)\omega'(b) = \beta(a, b)\omega'(ab) \quad (a, b \in G)$$

as claimed above.

For later use we observe that

1) We have for any $a, b \in G$, $m, n \in M, \omega(am, bn) = \omega(a, b)$.

2) Since $\beta(a, b) = \eta(m(s(a) \cdot s(b)))$ the map $\beta : G \times G \to T$ is smooth.

3) Writing ι for the canonical morphism $G \to G/M = K$ (say) there exists an $\alpha \in Z^2(K, T)$ such that $\beta = (\iota(\alpha))^{-1}$.

f) We continue now with the proof that if G is connected, simply connected and locally algebraic, then it is of type I. We start by noting that $K = G/M$ is reductive. Let us recall next (cf. [**AM**], Theorem 7, p. 61) that there is an α-factor representation of K, μ (say), such that $\lambda = (\omega' \otimes \iota(\mu))^{\sim}$, and the type of λ is the same as that of μ. We define a group K_a on the set $\{(u, \tau); u \in K, \tau \in T\}$ with the law of composition $(u, \tau)(v, \tau') = (uv, \alpha(u, v)\tau \cdot \tau')$. This is a connected Lie group, central extension of K by a 1-torus. μ is identifiable to a representation of K_a which on T restricts to a multiple of the identity map of T onto itself.

f1) Let $\mathfrak{k}$ be the Lie algebra of K, $\mathfrak{k}_a$ that of K_a, and $\mathfrak{t}$ the subalgebra corresponding to T of $\mathfrak{k}_a$. We write ψ for the canonical morphism $\mathfrak{k}_a \to \mathfrak{k}_a/\mathfrak{t}$.

Let $\mathfrak{i}$ be a Levi section in $\mathfrak{k}_a$. We note that $\mathfrak{i}+\mathfrak{t}$ is invariant under $\operatorname{ad}(\mathfrak{i})$. Since $\mathfrak{i}$ is semisimple, it can be complemented by an $\operatorname{ad}(\mathfrak{i})$-invariant subspace $\mathfrak{u}$ (say). Observe that the restriction of ψ to $\mathfrak{i}+\mathfrak{u}$ is an isomorphism with its image. We have $[\mathfrak{i}, \mathfrak{u}] \subseteqq \mathfrak{t}$, but also $[\mathfrak{i}, \mathfrak{u}] \subseteqq \mathfrak{u}$. We conclude from this that $[\mathfrak{i}, \mathfrak{u} + \mathfrak{t}] = 0$ and that $[\mathfrak{u}, \mathfrak{u}] \subseteqq \mathfrak{t}$. Let c be a nonzero element in $\mathfrak{t}$; we can write for $x, y \in \mathfrak{u} : [x, y] = B(x, y)c$. Let R be the radical of B, and V a complementary subspace to R. We write $\mathfrak{b} = V + \mathfrak{t}$; we have $\mathfrak{b} = (H)$. We conclude from all of this that $\mathfrak{k}_a = \mathfrak{i} + \mathfrak{b} + R$ (direct sum of Lie algebras) such that $\mathfrak{i}$ is semisimple, $\mathfrak{b}$ is Heisenberg, and R is central abelian. We write $\mathfrak{k}_a^c = \mathfrak{t} + R$.

f2) We derive from all this that we can write $K_a = D \times A_a \times A$ (direct product of Lie groups) where D is connected, simply connected semisimple; A_a is Heisenberg, divided by a discrete subgroup of the center, and A is a vector group. Let ρ be a factor representation of K_a, such that $\rho|T$ is quasi-equivalent to a multiple of the identity map of T onto itself. To show that ρ is of type I, we note that the space $\mathfrak{h}$ of ρ can be written as $\mathfrak{h}_1 \otimes \mathfrak{h}_2$, such that $\rho|(D \times A_a) = \rho_1 \otimes \rho_2$, where ρ_1 is a multiple of an irreducible representation of D, and ρ_2 that of $A_a \times A$.

This, however, clearly implies that ρ is of type I.

1.3. Proof of Main Lemma

This section is devoted to the proof of the

MAIN LEMMA. *Let $\mathfrak{g}$ and $\mathfrak{h}$ be algebraic Lie algebras on a finite-dimensional real vector space. We assume that $\mathfrak{h}$ is an ideal in $\mathfrak{g}$. Let G and H be connected and simply connected Lie groups belonging to $\mathfrak{g}$ and $\mathfrak{h}$ respectively; we assume that $G \supset H$. Then $\widehat{H}/G$ is countably separated.*

This proposition will play an important role in the subsequent section and in Chapter III. Its demonstration will require repetitions, in altered form, of some parts of the previous section.

1.3.1. First we recall certain facts of the representation theory of locally compact groups; details can be found, e.g., in [**D***].

Let N be a separable locally compact group. Given a unitary representation T of N, there is a subset $S(T)$ of $\widehat{N}$, well-determined by the condition that it be closed and weakly equivalent to T. If $N = (I)$, $S(T)$ is the closed hull of a canonical measure corresponding to T on $\widehat{N}$. Let H be a separable, locally compact group, containing N (as above) as a closed invariant subgroup. Given a Borel subset A of $\widehat{N}$, we shall write $\widehat{H}_A$ for the collection of all those irreducible unitary representations of H, the restriction of which to N is carried by A.

LEMMA 1. *Let A be a subset of $\widehat{N}$, which is locally closed and invariant with respect to H. Then $\widehat{H}_A$ is locally closed in $\widehat{H}$.*

In fact,

(i) We start by proving the following statement. Let $A \subset \widehat{N}$ be closed; then $\widehat{H}_A$ is closed in $\widehat{H}$. In fact, we have to prove the following. Assume that π is weakly contained in $\widehat{H}_A$. Then π belongs to $\widehat{H}_A$.

a) Our assumption implies that $\pi|N$ is weakly contained in $\widehat{H}_A|N$.

b) Assume that λ is an element of $\widehat{H}_A$. Then by definition, $\lambda|N$ is carried by A whence, since A is closed, $S(\lambda|N)$ is a subset of A.

Since $\lambda|N$ is weakly equivalent to $S(\lambda|N)$, $\lambda|N$ is weakly contained in A, and thus $\ker(\lambda|N)$ contains $\ker(A)$.

c) Since (cf. a)) $\pi|N$ is weakly contained in $\widehat{H}_A|N$, we have

$$\ker(\pi|N) \supseteqq \bigcap_{\lambda \in \widehat{H}_A} \ker(\lambda|N) \supseteqq \ker(A).$$

d) Since $\pi|N$ is weakly equivalent to $S(\pi|N)$, by c), $S(\pi|N) \subseteqq A$, and thus $\pi|N$ is carried by A. Hence we conclude that if π is weakly contained in $\widehat{H}_A$, then it belongs to $\widehat{H}_A$. In this manner, $\widehat{H}_A$ is closed in $\widehat{H}$.

(ii) We claim that if A is open and H-invariant in $\widehat{N}$, then $\widehat{H}_A$ is open in $\widehat{H}$. In fact, let λ be an element of $\widehat{H}$. By the H-invariance of A in $\widehat{N}$, $\lambda|N$ is carried either by A or by $\widehat{N} - A$. Hence $\widehat{H} - \widehat{H}_A$ is the set of all those $\lambda \in \widehat{H}$, for which $\lambda|N$ is concentrated on $\widehat{N} - A$, and hence, by (i), is closed in $\widehat{H}$.

(iii) We come now to the demonstration of the statement of Lemma 1. Below we are going to use the fact that by $(h, \pi) \mapsto h\pi$ $(h \in H, \pi \in \widehat{N})$, H acts as a topological transformation group on $\widehat{N}$ (cf. [**D***], 3.9.9, p. 81). We have to prove that if A is locally closed and H-invariant in $\widehat{N}$, then $\widehat{H}_A$ is locally closed in $\widehat{H}$. In fact, our assumption implies that we can write $A = F \cap O$, where F is closed and O is open in $\widehat{N}$. We have $A \subseteqq F$ and hence, since F is closed, $\overline{A} \subseteqq F$. We can thus assume that F is closed and H-invariant. Since A and F are H-invariant, we conclude that

$$A = F \cap O = F \cap \left(\bigcup_{h \in H} h(O) \right).$$

Summing up so far, we can assume that $A = F \cap O$, where each component is H-invariant. We conclude from here that by (i), $\widehat{H}_F$ is closed, by (ii) $\widehat{H}_O$ is open. In this fashion it suffices to note the following obvious statement. Let A and B be Borel in $\widehat{N}$. Then we have

$$\widehat{H}_A \cap \widehat{H}_B = \widehat{H}_{A \cap B}.$$

Summing up, we have completed the proof of Lemma 1: If A is locally closed and H-invariant in $\widehat{N}$, then $\widehat{H}_A$ is locally closed in $\widehat{H}$.

Let ω be given, fixed in $\widehat{N}$. We write $S = H_\omega$ and note that since N is of type I, S is closed in H. We have $N \subseteqq S \subseteqq H$. We set

$$\widehat{S}_\omega = \{\eta; \eta \in \widehat{S} \text{ such that } \eta|N \text{ is quasi-equivalent to } \omega\}.$$

LEMMA 2. *Unexplained notation as above. We claim that* $\widehat{S}_\omega$ *is locally closed in* $\widehat{S}$.

In fact, by $N = (I)$ there are closed 2-sided ideals $J \supset I$ in $C^*(N)$ such that $\omega \in \widehat{J} - \widehat{I}$ and the right-hand side is locally compact, Hausdorff (cf. [**D***], p. 94). We conclude from this that

$$\{\omega\} = \overline{\{\omega\}} \cap (\widehat{J} - \widehat{I})$$

or that $\{\omega\}$ is locally closed in $\widehat{N}$.

Since $s\omega = \omega$ $(s \in S)$ we infer by Lemma 1 that $\widehat{S}_\omega$ is locally closed in $\widehat{S}$, as claimed above.

Below we assume that H is of type I.

LEMMA 3. *Unexplained notation as above. We claim that the subset* $\widehat{S}_\omega$ *of* $\widehat{S}$, *with respect to the Borel structure induced by the topology of* $\widehat{S}$, *is standard.*

In fact,

(i) By Lemma 2, $\widehat{S}_\omega$ is locally closed in $\widehat{S}$, and hence there are closed, 2-sided ideals $J \supset I$ in $C^*(S)$ such that $\widehat{S}_\omega = \widehat{J} - \widehat{I} = (J/I)\widehat{}$.

(ii) Let us assume that $\widehat{S}_\omega$ is not standard. Then (cf. [**D***], 9.5.6, p. 185) J/I is not of type I, and thus there is a factor representation τ of J/I that is not of type I. We write τ' for the corresponding representation of $C^*(S)$ and σ for the corresponding representation of S. These are factor representations, none of which is of type I. We claim that $\sigma|N$ is a multiple of ω. In fact, we can form the decomposition into irreducibles

$$\tau = \int_X^\oplus \tau(x)\,.d\mu(x).$$

Assume that $\sigma(x)$, in $\widehat{S}_\omega$, is of the class of $\tau(x)$. We have thus

$$\sigma = \int_X^\oplus \sigma(x).\,d\mu(x)$$

from where we conclude that

$$\sigma|N = \int_X^\oplus (\sigma(x)|N).\,d\mu(x).$$

Since $\sigma(x)|N$ is a multiple of ω, the same is valid for $\sigma|N$. We conclude from all this that if $\widehat{S}_\omega$ is not standard, then there is a factor representation σ of S, which is not of type I, and for which $\sigma|N \approx \omega$. But then (cf. [**AM**], Theorem 6, p. 58) $\mathrm{ind}_{S\uparrow H}\,\sigma$, too, is a factor representation, which is not of type I, contradicting the assumption that H is of type I. Summing up, we have completed proving that, for any $\omega \in \widehat{N}$, $\widehat{S}_\omega$ is standard with respect to the Borel structure, induced by the topology of $\widehat{S}$.

LEMMA 4. *Let us write for* $\omega \in \widehat{N}$, *fixed,* $O = H\omega \subset \widehat{N}$. *Then* O *is a Borel subset of* $\widehat{N}$.

In fact, we shall make use of the following theorem of Souslin (cf. [**AM**], Proposition 2.5, p. 7). Let us assume that E is an analytic Borel space and that F is a countably generated Borel space. Assume also that $f : E \to F$ is a Borel injection. Then f is a Borel isomorphism of E onto $f(E)$; if E is standard, then $f(E)$ is Borel in F. To prove our statement, apply the above reminder to $E = H/H_\omega$, $F = \widehat{N}$ (both standard) and the map $hH_\omega \mapsto h\omega$ $(h \in H)$.

We recall that if F is standard and $E \subsetneqq F$ is Borel, then the Borel structure, induced by F on E, is again standard.

Below we assume that G is a separable locally compact group, containing H and N, as above, as closed, invariant subgroups.

LEMMA 5. *We claim that $\widehat{H}_O$ is Borel in $\widehat{H}$ and that there is a G_ω-equivariant Borel isomorphism $\widehat{S}_\omega \to \widehat{H}_O$.*

With the notation of Lemma 3, since $\widehat{S}_\omega$ is standard, by [**D***] 4.6.2, p. 85, there is a Borel map $\zeta \mapsto \pi(\zeta)$ from $\widehat{S}_\omega$ into $\mathrm{Irr}(S)$ such that $\pi(\zeta)$ is of the class of ζ.

We define

$$\delta(\zeta) = \mathrm{ind}_{S\uparrow H}\, \pi(\zeta).$$

Then (cf. [**Mackann I**], Theorem 10.1, p. 123) the map $\zeta \mapsto \delta(\zeta)$ from $\widehat{S}_\omega$ into $\mathrm{Irr}(S)$ is Borel such that, if $\sigma : \mathrm{Irr}(H) \to \widehat{H}$ is the canonical projection, then $\psi = \sigma \circ \delta$ is a Borel injection from $\widehat{S}_\omega$ onto its image in $\widehat{H}$, which is $\widehat{H}_O$. Hence, by the theorem of Souslin, $\widehat{H}_O$ is Borel. The G_ω-equivariance is implied by elementary functorial properties of the induction.

Below we assume that $\widehat{N}/G$ is countably separated.

LEMMA 6. *Unexplained notation as above. We write $\Omega = G\omega \subset \widehat{N}$ and claim that $\widehat{H}_\Omega$, in the Borel structure induced by that of $\widehat{H}$, is a standard Borel space.*

In fact, G acts on $\widehat{N}$ as a topological transformation group. Since N is of type I, $\widehat{N}$ is almost Hausdorff. We assume that $\widehat{N}/G$ is countably separated. Thus, by 1.1, Ω is locally closed and thus Borel in $\widehat{H}$. Since H is of type I, $\widehat{H}$ is standard, and hence so is $\widehat{H}_\Omega$.

LEMMA 7. *Unexplained notation as above. We shall also assume that if $\tau \in \widehat{H}_\Omega$, then $\tau|N$ gives rise to a transitive H-orbit on $\widehat{N}$. We claim that if one of the Borel spaces $\widehat{H}_\Omega/G$ and $\widehat{S}_\omega/G_\omega$ is countably separated, then so is the other and the two are Borel isomorphic.*

In fact, we start by noting that by Lemma 5, we can replace $\widehat{S}_\omega/G_\omega$ by $\widehat{H}_O/G_\omega$. Below we shall make a repeated use of the following proposition (cf. [**AM**], Proposition 2.11, p. 8). Assume that E is analytic and that F is countably separated. Let the map $f : E \to F$ be Borel. We define an equivalence relation on E such that $x \sim y$ $(x, y \in E)$ if and only if $f(x) = f(y)$. We denote again by f this equivalence relation. Then the natural map $E/f \to F$ is a Borel isomorphism from E/f onto $f(E)$, and E/f is analytic. This having been said, let us assume first that $\widehat{H}_O/G_\omega$ is countably separated. We propose to show that then $\widehat{H}_\Omega/G$ is countably separated.

(i) We observe that by virtue of our assumptions, the map $(a, \lambda) \mapsto a\lambda$ from $G \times \widehat{H}_\Omega$ into $\widehat{H}_\Omega$ is surjective. We shall denote it by T; it is clearly Borel.

(ii) Let us write ρ for the canonical morphism $\widehat{H}_O \to \widehat{H}_O/G_\omega$. We observe that there is a map $\phi : \widehat{H}_\Omega \to \widehat{H}_O/G_\omega$, well-determined by $\phi(a\lambda) = \rho(\lambda)$ $(a \in G, \lambda \in \widehat{H}_O)$. In fact, assume that $a_1\lambda_1 = a_2\lambda_2$. We have then $a_1H\omega = a_2H\omega$. Hence there is $h \in H, b \in G_\omega$ such that $a_1 = a_2bh$. Therefore $b\lambda_1 = \lambda_2$, whence $\rho(\lambda_1) = \rho(\lambda_2)$.

(iii) We observe that the map $\phi : \widehat{H}_\Omega \to \widehat{H}_O/G_\omega$ is Borel. In fact,

a) We define $\varphi : G \times \widehat{H}_O \to \widehat{H}_O/G_\omega$ by $\varphi((a,\lambda)) = \rho(\lambda)$; clearly it is Borel. We have

$$\begin{array}{ccc} G \times \widehat{H}_O & \xrightarrow{T} & \widehat{H}_\Omega \\ & \searrow^{\varphi} & \downarrow{\scriptstyle \phi} \\ & & \widehat{H}_O/G_\omega \end{array}$$

b) Assuming that B is a Borel set in $\widehat{H}_O/G_\omega$ we write

$$D = \phi^{-1}(B).$$

We must show that D is Borel in $\widehat{H}_\Omega$. We have that

$$\varphi^{-1}(B) = T^{-1}(D)$$

is Borel. Lemma 5 implies that $\widehat{H}_O$ is Borel in $\widehat{H}$, and thus $E = G \times \widehat{H}_O$ is standard. By Lemma 6, $\widehat{H}_\Omega$ $(= F$, say) is standard. We apply now the reminder above with E, F, T as just specified. The conclusion we derive is that $\widehat{H}_\Omega = (G \times \widehat{H}_O)/T$ and hence $D = T(\varphi^{-1}(B))$ is Borel. Summing up, we have proved that the map

$$\phi : \widehat{H}_\Omega \to \widehat{H}_O/G_\omega$$

is Borel. We recall that our objective is that $\widehat{H}_\Omega/G$ is Borel isomorphic to $\widehat{H}_O/G_\omega$ (if $\widehat{H}_O/G_\omega$ is countably separated).

(iv) We apply once more the reminder at the start of this lemma as follows. We take $E = \widehat{H}_\Omega, F = \widehat{H}_O/G_\omega$, and $f = \phi$. We conclude from it that

$$\widehat{H}_\Omega/\phi = \widehat{H}_O/G_\omega.$$

In this manner it is enough to prove that on $\widehat{H}_\Omega$: $\{\phi\text{-fibers}\} = \{G\text{-orbits}\}$. In fact, let μ_1, μ_2 in $\widehat{H}_\Omega$ be such that $\phi(\mu_1) = \phi(\mu_2)$. Let us write $\mu_k = a_k\lambda_k$ $(a_k \in G, \lambda_k \in \widehat{H}_O; k = 1,2)$. We conclude that

$$\rho(\lambda_1) = \phi(\mu_1) = \phi(\mu_2) = \rho(\lambda_2)$$

and therefore there exists a b in G_ω such that $\lambda_2 = b\lambda_1$ or that $G\mu_1 = G\mu_2$. By the same token $G\mu_1 = G\mu_2$ implies $\phi(\mu_1) = \phi(\mu_2)$. Summing up, we have completed proving that if $\widehat{H}_O/G_\omega$ is countably separated, then so is $\widehat{H}_\Omega/G$ and the two are Borel isomorphic. Let us assume now that $\widehat{H}_\Omega/G$ is countably separated. We are going to show that $\widehat{H}_O/G_\omega$ is countably separated and the two are Borel isomorphic. Let I be the canonical injection $\widehat{H}_O \to \widehat{H}_\Omega$ and π the canonical projection

$$\widehat{H}_\Omega \to \widehat{H}_\Omega/G.$$

Let us put $\psi = \pi \circ I$. We have evidently $\widehat{H}_\Omega/G \to \widehat{H}_O/\psi$; hence it suffices to note that if λ_1, λ_2 in $\widehat{H}_O$ are such that $\psi(\lambda_1) = \psi(\lambda_2)$, then $G\lambda_1 = G\lambda_2$ and thus $\lambda_2 = a\lambda_1$ for some $a \in G$. The converse is evident. We sum up the previous discussion as follows. Assume that any $\tau \in \widehat{H}_\Omega$ restricts to a transitive H-orbit. Then we have the following situation. If any of the two spaces $\widehat{S}_\omega/G_\omega$ or $\widehat{H}_\Omega/G$ is countably separated, then so is the other, and the two are Borel isomorphic.

LEMMA 8. *Let G, H, N be as above. Then $\widehat{H}/G$ is countably separated if and only if for each $\omega \in \widehat{N}$, $\widehat{S}_\omega/G_\omega$ is countably separated.*

In fact,

a) Assume first that $\widehat{H}/G$ is countably separated. Let ω be an element of $\widehat{N}$; we shall show that $\widehat{S}_\omega/G_\omega$ is countably separated.

Let us put $\Omega = G\omega$. Since $\widehat{N}/G$ is countably separated, Ω is locally closed in $\widehat{N}$ (cf. 1.1). By Lemma 1 above this implies that $\widehat{H}_\Omega$ is locally closed in $\widehat{H}$. Hence, if $\widehat{H}/G$ is countably separated, then $\widehat{H}_\Omega/G$, too, is countably separated.

b) We have to show that if $\widehat{S}_\omega/G_\omega$ is countably separated for all $\omega \in \widehat{N}$, then $\widehat{H}/G$ is countably separated. Below we shall make much use of 1.1.

Since $H = (I)$, $\widehat{H}$ is almost Hausdorff. Hence $\widehat{H}/G$ is countably separated if and only if, for all λ in $\widehat{H}$, $H\lambda$ is locally closed. Let $\omega \in \widehat{N}$ be such that $\lambda|N$ is carried by $G\omega = \Omega$, say. Ω is H-invariant and locally closed, hence (by Lemma 1 above) $\widehat{H}_\Omega$ is locally closed in $\widehat{H}$. Our assumption is that $\widehat{S}_\omega/G_\omega$ is countably separated; hence, by Lemma 7, so is $\widehat{H}_\Omega/G$. But, by $H = (I)$ and $\widehat{H}_\Omega$ is locally closed, $\widehat{H}_\Omega$ is almost Hausdorff. From here we conclude that $H\lambda$ is locally closed in $\widehat{H}_\Omega$; hence, finally, $H\lambda$ is locally closed in $\widehat{H}$.

1.3.2. Below, notions concerning $\mathfrak{g}$, analogous to those with respect to $\mathfrak{h}$, will be distinguished by an index zero.

Let $\mathfrak{m}$ be the greatest ideal in $\mathfrak{h}$, composed of nilpotent endomorphisms only. Let $\mathfrak{r}$ be the radical of $\mathfrak{h}$. We recall (cf. [**B**], Corollaire 6, p. 67) that if l is a nilpotent endomorphism on V belonging to $\mathfrak{r}_0$, then l lies in $\mathfrak{m}_0$.

LEMMA 1. *We have* $\mathfrak{m} = \mathfrak{m}_0 \cap \mathfrak{h}$.

In fact, assume that l is in $\mathfrak{m}$; then l is nilpotent, and hence $\mathfrak{m}$ is a nilpotent ideal of $\mathfrak{h}$. This implies that

$$\mathfrak{m} \subseteqq \mathfrak{r} = \mathfrak{h} \cap \mathfrak{m}_0 \subseteqq \mathfrak{r}_0$$

and thus $\mathfrak{m} \subseteqq \mathfrak{r}_0$. From this, using the reminder above we conclude that $\mathfrak{m} \subseteqq \mathfrak{m}_0$ which gives $\mathfrak{m} \subseteqq \mathfrak{m}_0 \cap \mathfrak{h}$.

Conversely, we first observe that $\mathfrak{m}_0 \cap \mathfrak{h}$ is an ideal in $\mathfrak{h}$ and, if l is in $\mathfrak{m}_0 \cap \mathfrak{h}$, then l is nilpotent, whence $\mathfrak{m}_0 \cap \mathfrak{h} \subseteqq \mathfrak{m}$.

Summing up, we have proved that $\mathfrak{m} = \mathfrak{m}_0 \cap \mathfrak{h}$ and hence, in particular, $\mathfrak{m}$ is an ideal in $\mathfrak{g}$.

LEMMA 2. *Unexplained notation as above. We claim that the principal statement is valid, if* $\mathfrak{m} = \{0\}$.

In fact, if so, then $\mathfrak{h}$ is reductive and hence $\mathfrak{h} = \mathfrak{h}_1 \times \mathfrak{h}_2$, where $\mathfrak{h}_1 = [\mathfrak{h}, \mathfrak{h}]$ is semi-simple and $\mathfrak{h}_2 = \mathfrak{h}^{\mathrm{c}}$.

Writing H_k for the connected and simply connected Lie group corresponding to $\mathfrak{h}_k$ $(k = 1, 2)$, we have

$$H = H_1 \times H_2 \quad \text{and} \quad \widehat{H} = \widehat{H}_1 \times \widehat{H}_2.$$

Since $\mathfrak{h}_1$ is semi-simple, we conclude that $\mathrm{Ad}(G)|\mathfrak{h}_1 = \mathrm{Ad}(H_1)|\mathfrak{h}$, and hence $\widehat{H}/G = \widehat{H}_1 \times (\widehat{H}_2/G)$. Since $\mathfrak{g}$ is algebraic, $\mathrm{Ad}(G)/\mathfrak{h}_2$ is the connected component of the identity of a linear algebraic group in $\mathrm{GL}(\mathfrak{h}_2)$ and hence, by the theorem of Chevalley, $\widehat{H}_2/G$ is countably separated. Summing up, we have completed proving the following: If $\mathfrak{h}$ contains no nontrivial ideal, composed of nilpotent endomorphisms, then $\widehat{H}/G$ is countably separated.

Below we shall assume $\mathfrak{m} \neq \{0\}$. To prove the main result, we shall proceed by induction; we shall assume its validity for pairs $(\mathfrak{g}', \mathfrak{h}')$ satisfying $\dim(\mathfrak{g}') < \dim(\mathfrak{h}')$. We recall (cf. 1.1) that since H is of type I (cf. Theorem in 1.2) $\widehat{H}$ is almost Hausdorff and hence $\widehat{H}/G$ is countably separated, if and only if for any λ in $\widehat{H}$, $G\lambda$ is locally closed in $\widehat{H}$. We note that there is an abelian nonzero $\mathfrak{G}$-ideal in $\mathfrak{m}$ (e.g. $\mathfrak{m}^c$). Let $\mathfrak{a}$ be such. We denote by A the closed, invariant, connected subgroup of H and G, corresponding to $\mathfrak{a}$. We note (cf. 1.1) that by the theorem of Chevalley, $\widehat{A}/H$ and $\widehat{A}/G$ are countably separated, and thus $\lambda|A$ is carried by an H-orbit, $H\omega$, say $(\omega \in \widehat{A})$. We write $\Omega = G\omega$. Ω is G-invariant and locally closed in $\widehat{A}$.

LEMMA 3. *Writing $S = H_\omega$, S is of type* I.

In fact, this was already proved in 1.2 (cf. (iii) e) loc. cit.; note that now $S_0 = (I)$ can be derived from the main theorem loc. cit.).

To show that $G\lambda$ is locally closed, it is enough to establish that if $\lambda|A$ is carried by $\Omega = G\omega \subset \widehat{A}$, then $\widehat{H}_\Omega/G$ is countably separated. By Lemma 7 in 1.3.1 this is so if and only if $\widehat{S}_\omega/G_\omega$ is countably separated. (Recall that $\widehat{S}_\omega = \{\mu; \mu \in \widehat{S}$ such that $\mu|A \approx \omega$; $\widehat{S}_\omega$ is, by Lemma 2 in 1.3.1, locally closed in $\widehat{S}$.)

We shall distinguish two cases.

A) Either $\dim(\mathfrak{a}) > 1$ or $\dim(\mathfrak{a}) = 1$, but $\mathfrak{a}$ is not contained in $\mathfrak{g}^c$.

B) The conditions of A) cannot be realized.

A) Here we shall distinguish two subcases:

A1) $\dim(\Omega) > 0$; A2) $\dim(\Omega) = 0$.

A1) To obtain the desired conclusion, here it is enough to establish

LEMMA 4. *Assume that* $\dim(G_\omega) < \dim(G)$. *Then* $\widehat{S}/G_\omega$ *is countably separated.*

In fact,

(i) We write $F = (G_\omega)_0$ and claim that $\widehat{S}_0/F$ is countably separated. Let $G_1 \subset \mathrm{GL}(V)$ be the irreducible algebraic group corresponding to the Lie algebra $\mathfrak{g}$. Then we have for any $a \in \mathfrak{a}, g \in G_1 : gag^{-1} \in \mathfrak{a}$. G_1 operates rationally on $\widehat{A}$, and hence $(G_1)_\omega$ is algebraic. Let us denote by $\mathfrak{g}_\omega$ its Lie algebra; it is algebraic. Similarly, $\mathfrak{h}_\omega$ is algebraic; it is an ideal in $\mathfrak{g}_\omega$. We write $(\mathcal{F}, p)$ for the universal covering of F (defined as $(G_\omega)_0$).

a) Let us write γ for the connected subgroup, corresponding to $\mathfrak{h}_\omega$ in $\mathcal{F}$. Since $\mathfrak{h}_\omega$ is an ideal in $\mathfrak{g}_\omega$, γ is closed, invariant and simply connected. We write q for $p|\gamma$; (γ, q) is the universal covering of S_0.

b) By virtue of the assumption of our inductive procedure, $\widehat{\gamma}/\mathcal{F}$ is countably separated.

c) Given $\pi \in \widehat{S}_0$, we can form $(q^*\pi)(x) = \pi(q(x))$ $(x \in \gamma)$.

1) $\pi \mapsto q^*\pi$ is a continuous bijection $\widehat{S}_0 \to q^*(\widehat{S}_0)$. We write $T = q^*(\widehat{S}_0)$.

2) Let us put $\mathcal{J} = \ker(q)$. Then $\sigma \in \widehat{\gamma}$ will belong to T if and only if $\sigma|\mathcal{J}$ is identically one. Hence T is a closed, $\mathcal{F}$-invariant subset of $\widehat{\gamma}$.

3) We observe that the map $q^* : \widehat{S}_0 \to T$ makes correspond to each other F and $\mathcal{F}$-orbits respectively. In fact, let us choose π in $\widehat{S}_0$ and put $\pi' = q^*\pi$. We have on γ,

$$\begin{aligned}(a\pi')(x) &= \pi'(a^{-1}xa) = \pi(q(a^{-1}xa)) \\ &= \pi((p(a))^{-1}p(x)p(a)) = (p(a)\pi)'(x) \quad (x \in \gamma).\end{aligned}$$

Conclusion. The map q^* makes correspond to each other $T/\mathcal{F}$ and $\widehat{S}_0/F$.

4) By virtue of the hypothesis of our induction, $\widehat{\gamma}/\mathcal{F}$ is countably separated. Since q^* is a continuous map from $\widehat{S}_0$ onto T, if B is Borel in T, then so is $(q^*)^{-1}(B)$ in $\widehat{S}_0$. Also, by 3), if B is $\mathcal{F}$-invariant in T, then $(q^*)^{-1}(B)$ is F-invariant in $\widehat{S}_0$. Summing up, by all we said so far, the induction assumption implies that $\widehat{S}_0/F$ is countably separated. Our goal is to show that $\widehat{S}/G_\omega$ is countably separated.

(ii) a) We observe that evidently there is a discrete subgroup L in G^c such that G_ω/LF is finite. We conclude from this that G_ω acts on $\widehat{S}_0/F$ as a finite group, Γ (say).

b) Next we note that $\widehat{S}_0/F$ is analytic. In fact, by what we saw in (i), $\widehat{S}_0/F$ is countably separated. We recall once more (cf. [**AM**], Proposition 2.11, p. 9) that if E is analytic, E' is countably separated, and $f : E \to E'$ is a Borel map, then E/f, via f, is Borel isomorphic to $f(E) \subset E'$, *and* E/f *is analytic.*

We apply this to $E = \widehat{S}_0$ (standard by $S_0 = (I)$), $E' = \widehat{S}_0/F$ (countably separated by (i)) and $s \mapsto Fs$ ($s \in \widehat{S}_0$), from where we conclude that $\widehat{S}_0/F$ is analytic.

c) We recall that by the proof of the proposition in 1.2, if (E, Γ) is an analytic Borel Γ-space and Γ is finite, then E/Γ is countably separated. We conclude from all this that $\widehat{S}_0/G_\omega$ is countably separated.

(iii) As we saw above, there is an $L \subset S \cap G^c$ such that if we put $S_1 = LS_0$, then S/S_1 is finite. We recall now that by Lemma 7 in 1.3.1, if G is separable, locally compact, $H \supset N$ are closed, invariant type I subgroups of G such that $\widehat{N}/G$ is countably separated, and for any $\tau \in \widehat{H}$, $\tau|N$ is transitive, then $\widehat{H}/G$ is countably separated if and only if for all ω in $\widehat{N}$, $\widehat{S}_\omega/G_\omega$ is countably separated ($\widehat{S}_\omega = \{\eta; \eta \in \widehat{S}$ such that $\eta|N \approx \omega\}$).

We conclude from this that $\widehat{S}_1/G_\omega$ is countably separated by replacing G by G_ω, H by S_1 and N by S_0. In fact,

1) $S_0 = (H_\omega)_0$ is locally algebraic and hence of type I.

2) Clearly $S_1 = (I)$.

3) S_1 acts trivially on $\widehat{S}_0$.

4) $\widehat{S}_0/G_\omega$ is countably separated by (ii).

Conclusion: Take $\eta \in \widehat{S}_0$ and set $\widehat{T}_\eta = \{\mu; \mu \in \widehat{S}_1,\ \mu|S_0 = \eta\}$. Then $(G_\omega)_\eta$ operates trivially on $\widehat{T}_\eta$. All this having been said, we can conclude by Lemma 7 of 1.3.1 that $\widehat{S}_1/G_\omega$ is countably separated.

(iv) We complete now the proof of Lemma 4 by another application of Lemma 7 in 1.3.1. In fact, take now G_ω for G, S for H, and S_1 for N. The only thing to be explained is that $\widehat{S}_1/G_\omega$ is countably separated but this has just been proved in (iii). It remains to be observed that $\widehat{S}_\omega/G_\omega$ is countably separated. But this is so, since $\widehat{S}_\omega$ is locally closed in $\widehat{S}$. We have just completed the proof of Lemma 4: If $\dim(G_\omega) < \dim(G)$, then $\widehat{S}/G_\omega$ is countably separated.

Before proceeding, we recall that our ultimate objective is to establish that $\widehat{H}/G$ is countably separated, or, what is the same, that for each $\lambda \in \widehat{H}$, $G\lambda$ is locally closed. To this end, we have already proved that we can assume that $\mathfrak{m} \neq \{0\}$, where $\mathfrak{m}$ is the greatest ideal composed only of nilpotent endomorphisms (cf. Lemma 2).

Next we distinguish two cases; namely,

A) There is a $\mathfrak{g}$-ideal $\mathfrak{a}$ in $\mathfrak{m}$ such that either $\dim(\mathfrak{a}) > 1$ or $\dim(\mathfrak{a}) = 1$ but $\mathfrak{a}$ is not contained in the center of $\mathfrak{g}$.

B) The condition of A) cannot be realized.

Assume now that we are in case A). Let λ be an element of $\widehat{H}$, giving rise to the G-orbit Ω on $\widehat{A}$. We have just shown (cf. Lemma 4) that in case A1) if $\dim(\Omega) > 0$, then $G\lambda$ is locally closed in $\widehat{H}$. Inside A) we still have to consider the case when $\dim(\Omega) = 0$ (Case A2)). The discussion of this possibility requires only a mechanical repetition of the analogous steps in 1.2), to which the reader is referred for more details.

A2) Let ω be an element in Ω. Then ω is G-invariant such that $\lambda|A \approx \omega$. We put $J = (\ker \omega)_0$; the dimension of J is positive. Let $j = \mathrm{Lie}(J)$. There is a rational representation ρ of $\mathfrak{g}$ such that $\ker(\rho) = j$ and $\rho(\mathfrak{h}) = (a)$. This having been said, we note that $(\rho(\mathfrak{g}), \rho(\mathfrak{h}))$ is as $(\mathfrak{g}, \mathfrak{h})$, but $\dim(\rho(\mathfrak{g})) < \dim(\mathfrak{g})$.

Let $\mathfrak{G}$ be a connected and simply connected group corresponding to $\rho(\mathfrak{g})$ and let $\mathfrak{H}$ be the connected subgroup given rise to by $\rho(\mathfrak{h})$. We have canonically $\mathfrak{H} = H/J$, and $\hat{\mathfrak{H}}$ identifies with a closed subset of $\widehat{H}$. We also have a correspondence between $\hat{\mathfrak{H}}/\mathfrak{G}$ and $\widehat{H}/\mathfrak{G}$. Let $\varphi : H \to H/J$ be the canonical morphism. Then there exists $\tilde{\lambda} \in \hat{\mathfrak{H}}$ such that $\tilde{\lambda} \circ \varphi = \lambda$ (by $J \subseteqq \ker(\lambda)$) whence the conclusion is clear.

1.3.3. B) Here we assume that if $\mathfrak{a}$ is an abelian $\mathfrak{g}$-ideal in $\mathfrak{m}$, then $\dim(\mathfrak{a}) = 1$ and $\mathfrak{a} \subseteqq \mathfrak{g}^c$. We distinguish two subcases, according to whether: B1) $\mathfrak{m}$ is abelian; B2) $[\mathfrak{m}, \mathfrak{m}] \neq 0$.

B1) Our discussion here will be similar to what we have done in 1.3.2, Lemma 2. Our assumption implies that $\mathfrak{m} \subset \mathfrak{g}^c$ and $\dim(\mathfrak{m}) = 1$. Since $\mathfrak{h} = (a)$, by [**Ch III**], Proposition 5, p. 144 we can write $\mathfrak{h} = \mathfrak{i} + \mathfrak{a} + \mathfrak{m}$, where $\mathfrak{i}$ is semi-simple, and $\mathfrak{h}^c = \mathfrak{a} + \mathfrak{m}$. We put $\mathfrak{i} = \mathfrak{h}_1, \mathfrak{h}^c = \mathfrak{h}_2$. Let H_1 and H_2 be the connected subgroups in H corresponding to $\mathfrak{h}_1$ and $\mathfrak{h}_2$ respectively. Then $H = H_1 \times H_2$ and $\widehat{H} = \widehat{H}_1 \times \widehat{H}_2$, and also $\widehat{H}/G = \widehat{H}_1 \times \widehat{H}_2/G$. Next we observe that if $\mathfrak{g} = (a)$ on V, then $\mathrm{ad}(\mathfrak{g}) = (a)$ on $\mathrm{GL}(V)$. Hence, by the theorem of Chevalley, $\widehat{H}_2/G$ is countably separated, and thus the same holds for $\widehat{H}/G$, which has been our objective.

B2) Next we assume that $[\mathfrak{m}, \mathfrak{m}] \neq 0$. From this we conclude, as in 1.2 (v) e), that $\mathfrak{m} = (H)$. Let $\lambda \in \widehat{H}$ be given. We write M for the connected subgroup corresponding to $\mathfrak{m}$. Then $\lambda|M^c$ is a multiple of some $\eta \in \mathrm{X}(M^c)$. We can assume that η is not identically one. In fact, otherwise we can reason as in A2) in 1.3.2.

Since $M = (H)$ and $\eta \not\equiv 1$, there is an irreducible unitary representation of M (ω, say), well-determined up to unitary equivalence, by the condition that $\omega|M^c$ be a multiple of η (cf. [**N**]). By $M^c \subset G^c$, ω is G-invariant. We put $\zeta = \{\omega\} \subset \widehat{M}$; ζ is locally closed and H-invariant. Hence by Lemma 1 in 1.3.1, $\widehat{H}_\zeta$ is locally closed. Since ω is G-invariant, $\widehat{H}_\zeta$, too, is G-invariant. Evidently, λ belongs to $\widehat{H}_\zeta$. We conclude from this that it is enough to establish that $\widehat{H}_\zeta/G$ is countably separated. In fact, if so, then by 1.1, $G\lambda$ is locally closed in $\widehat{H}_\zeta$. Since $\widehat{H}_\zeta$ is locally closed in $\widehat{H}$, $G\lambda$, too, will be shown to be locally closed in $\widehat{H}$.

LEMMA 1. *Let* $\mathfrak{u}_0$ ($\mathfrak{u}$) *be the radical of* $\mathfrak{g}$ ($\mathfrak{h}$ *resp.*). *Then there is an ideal* $\mathfrak{g}_1$ *in* $\mathfrak{g}$ *such that* $\mathfrak{g} = \mathfrak{g}_1 + \mathfrak{h}$ (*sum of spaces*) *and* $[\mathfrak{g}_1, \mathfrak{h}] \subseteqq \mathfrak{m}$.

In fact, let $\mathfrak{i}$ and $\mathfrak{i}_2$ be Levi subalgebras in $\mathfrak{g}$ and $\mathfrak{h}$ respectively. We can assume that $\mathfrak{i}_2 = \mathfrak{i} \cap \mathfrak{h}$ (cf. [**B**] Corollaire 4, p. 81). Hence there is an ideal $\mathfrak{i}_1$ in $\mathfrak{i}$ such that

$\mathfrak{i} = \mathfrak{i}_1 + \mathfrak{i}_2$ (direct sum of Lie algebras). We define $\mathfrak{g}_1$ by $\mathfrak{i}_1 + \mathfrak{u}_0$. This being so, we have

1) $\mathfrak{g}_1 + \mathfrak{h} = (\mathfrak{i} + \mathfrak{u}_0) + (\mathfrak{i}_2 + \mathfrak{u}) = \mathfrak{i} + \mathfrak{u}_0 = \mathfrak{g}$.

2) Next we note that $\mathfrak{g}_1$ is an ideal in $\mathfrak{g}$. We have, in fact, $[\mathfrak{g}_1, \mathfrak{g}] = [\mathfrak{i}_1 + \mathfrak{u}_0, \mathfrak{i} + \mathfrak{u}_0] \subseteqq [\mathfrak{i}_1, \mathfrak{i}] + \mathfrak{u}_0 \subseteqq \mathfrak{i}_1 + \mathfrak{u}_0 = \mathfrak{g}_1$ or that $[\mathfrak{g}_1, \mathfrak{g}] \subseteqq \mathfrak{g}_1$. Thus $\mathfrak{g}_1$ is an ideal in $\mathfrak{g}$.

3) We claim that $[\mathfrak{g}_1, \mathfrak{h}] \subseteq \mathfrak{h} \cap [\mathfrak{g}, \mathfrak{u}_0]$. In fact we have $[\mathfrak{g}_1, \mathfrak{h}] = [\mathfrak{i}_1 + \mathfrak{u}_0, \mathfrak{i}_2 + \mathfrak{u}] \subseteqq [\mathfrak{u}_0, \mathfrak{i}_2] + [\mathfrak{i}_1, \mathfrak{u}] + [\mathfrak{u}_0, \mathfrak{u}] \subseteqq [\mathfrak{g}, \mathfrak{u}_0]$; hence

$$[\mathfrak{g}_1, \mathfrak{h}] \subseteqq \mathfrak{h} \cap [\mathfrak{g}, \mathfrak{u}_0].$$

4) We recall (cf. [**B**], p. 65) that the nilpotent radical of a Lie algebra ($= \mathfrak{S}$, say) is the intersection of the kernels of finite-dimensional simple representations. We have (cf. loc. cit.) $\mathfrak{S} = [\mathfrak{g}, \mathfrak{g}] \cap \mathfrak{u}_0$. Next we observe that, since $\mathfrak{m}_0$ is the greatest ideal of nilpotency of the identity representation of $\mathfrak{g}$ on V, we have $[\mathfrak{g}, \mathfrak{u}_0] \subseteqq [\mathfrak{g}, \mathfrak{g}] \cap \mathfrak{u}_0 \subseteqq \mathfrak{m}_0$. In this fashion we obtain that

$$[\mathfrak{g}_1, \mathfrak{h}] \subset \mathfrak{h} \cap [\mathfrak{g}, \mathfrak{u}_0] \subseteqq \mathfrak{h} \cap \mathfrak{m}_0 \subseteqq \mathfrak{m}.$$

Summing up, we have proved that there is an ideal $\mathfrak{g}_1$ of $\mathfrak{g}$ such that $\mathfrak{g} = \mathfrak{g}_1 + \mathfrak{h}$ and $[\mathfrak{g}_1, \mathfrak{h}] \subseteqq \mathfrak{m}$.

LEMMA 2. *Let G_1 and H be the connected subgroups corresponding in G to $\mathfrak{G}_1$ and $\mathfrak{h}$, respectively. Then we have $G = G_1 H$ and if $r \in G_1$ and $h \in H$, then $rhr^{-1}.h^{-1} \in M$.*

In fact,

(i) Since $\mathfrak{g}_1$ and $\mathfrak{h}$ are ideals in $\mathfrak{g}$, $G_1 H$ is a subgroup in G. By $\mathfrak{g} = \mathfrak{g}_1 + \mathfrak{h}$, it is open and hence, since G is connected, $G_1 H = G$.

(ii) Next we show that if $r \in G_1$ and $h \in H$ are small, then $rhr^{-1}.h^{-1} \in M$. To this end, assume that $r = \exp(u)$ ($u \in \mathfrak{G}_1$) and $h = \exp(v)$ ($v \in \mathfrak{h}$). Then we have $rhr^{-1}.h^{-1} = \exp(\mathrm{ad}(r)v).\exp(-v)$. But $\mathrm{Ad}(r)v = \exp(\mathrm{ad}\, u)v = v + m$, where $m \in \mathfrak{m}$. Thus we get that $rhr^{-1}.h^{-1} = \exp(v+m).\exp(-v) \in M$ by the Hausdorff-Campbell formula. We conclude from this that if $r \in G_1$, $h \in H$ are small, then $rhr^{-1}.h^{-1} \in M$. In other words, r and h commute modulo M.

(iii) From here, since G_1 and H are connected, the same is true for arbitrary elements of G_1 and H.

Below we assume that $\mathfrak{m} = (H)$.

Before proceeding to the completion of the proof of the Main Lemma, we recall a few things, mostly from 1.2, which we have already discussed. Let us suppose that we are given an element ω of $\widehat{M}$, such that $\dim(\omega) > 1$, and such that $\omega(\exp(l)) = \exp(i(l,h))I$ ($l \in \mathfrak{m}^c$), $I =$ identity and h is not orthogonal to $\mathfrak{m}^c$. Then we have

$$Mh = h + (\mathfrak{m}^c)^{\perp}.$$

This implies at once that $G = G_h.M$. We write $K = H/N$ and $\iota : H \to K$ will be the canonical projection. We recall that if ω' and ω'' are projective extensions, belonging to the same cocycle, of ω from M to H, then there exists a $\chi \in \mathrm{X}(K)$ such that $\omega'' = \iota(\chi)\omega'$.

LEMMA 3. *With the above notation, there is a projective extension ω' of $\omega \in \widehat{M}$ to H, and a continuous element χ of $\mathrm{Hom}((G_1)_h, \mathrm{X}(K))$, such that*

$$a\omega' = \iota(\chi(a))\omega' \quad (a \in (G_1)_h).$$

In fact,

(i) The following results were established in 1.2.

1) There is a continuous unitary representation T of G_h on the space of ω, such that for any $a \in G_h$ and $m \in M$ we have

$$T(a)\omega(m)T(a^{-1}) = \omega(ama^{-1}).$$

2) Let us write $\pi : G \to G/M$ for the canonical projection. There is a smooth map $\sigma : G/M \to G_h$ such that $\pi(\sigma(s)) \equiv s$ $(s \in G/M)$. We write $s(a) \equiv \sigma(\pi(a))$ $(a \in G)$. Then there is a smooth map $m : G \to M$ such that

$$a = s(a)m(a) \quad (a \in G).$$

Note that if $a \in G_h$, then $m(a) \in M_h = M^{\mathrm{c}}$.

3) This being so, setting for $a \in H : \omega'(a) = T(s(a))\omega(m(a))$ and for $a, b \in H$

$$\beta(a,b) \equiv \overline{\eta(m(s(a)s(b)))}$$

we have $\omega'(a)\omega'(b) = \beta(a,b)\omega'(a).\omega'(b)$. Below we shall write α for $(\iota(\beta))^{-1}$.

(ii) We recall that by virtue of Lemma 2, if $r \in G_1$ and $h \in H$, then $[r,h] \in M$. If, furthermore, $r \in (G_1)_h$ and $h \in H_h$, then $[r,h] \in M^{\mathrm{c}}$. This being so, given $c \in (G_1)_h$ fixed, we can define

$$\varphi_c(a) \equiv \eta([(s(a))^{-1}, c^{-1}]).$$

Next we claim that for any $a \in H$, we have

$$(c\omega')(a) = \varphi_c(a).T(c^{-1})\omega'(a)T(c).$$

In fact, assume $a = rm$ $(r = s(a), m = m(a))$. We set $r_c = r^{-1}.c^{-1}.rc$, and hence we get $rr_c = c^{-1}rc$. In this manner, we can conclude that

$$\begin{aligned}(c\omega')(a) &= \omega'(c^{-1}ac) = T(r)\omega(r_c.c^{-1}mc)\\ &= \eta(r_c)T(r)\omega(c^{-1}mc) = \eta(r_c)T(r)T(c^{-1})\omega(m)T(c).\end{aligned}$$

We have $\eta(r_c) = \eta([r^{-1}, c^{-1}]) = \varphi_c(a)$ and $T|M^{\mathrm{c}} = 1$ from where we conclude that for $r \in H_h$ and $c \in (G_1)_h$

$$T(r) = T(c^{-1})T(r)T(c)$$

and hence, since $\omega'(a) = T(r)\omega(m)$, we obtain

$$(c\omega')(a) = \varphi_c(a)T(c^{-1})\omega'(a)T(c)$$

as claimed above.

(iii) a) For $c \in (G_1)_h$ fixed and $a \in H_h$, we write $\psi_c(a) \equiv \eta([c,a])$. Then, as a function of $a \in H_h$, this is an element of $\mathrm{X}(H_h)$. In fact, we have to show that if $a, b \in H_h$, then we have

$$\eta([c,ab]) = \eta([c,a])\eta([c,b]).$$

We note that if G is any group, and $a, b, c \in G$ then

$$[ab,c] = a[b,c]a^{-1}[a,c].$$

We conclude from this that if $c \in (G_1)_h$; $a, b \in H_h$, then $[b,c] \in M^{\mathrm{c}} \subset G^{\mathrm{c}}$ and thus $a[b,c]a^{-1} = [b,c]$, whence $\eta([c,ab]) = \eta([c,a])\eta([c,b])$ and $\psi_c(ab) = \psi_c(a)\psi_c(b)$.

b) We observe that $\psi_c|M^{\mathrm{c}} = 1$. In fact, assume that $a \in M^{\mathrm{c}}$. Then $\psi_c(a) = \eta([c,a]) = 1$ by $a \in M^{\mathrm{c}} \subset G^{\mathrm{c}}$. We conclude from this that by $M_h = H_h \cap M$ and $H = H_hM$, there exists $\psi'_c \in \mathrm{X}(H)$, well-determined through $\psi'_c|H_h = \psi_c, \psi_c|M^{\mathrm{c}} = 1$.

c) Assume that $a \in H_h$ is fixed. Then $(c \mapsto \psi'_c(a))$ is a morphism from $(G_1)_h$ into $\mathrm{X}(H)$.

In fact, we have always $[c_1c_2, a] = c_1[c_2, a]c_1^{-1}.[c_1, a]$. But by $c_2 \in (G_1)_h$ and $a \in H_h$ we have $[c_2, a] \in G^{\mathrm{c}}$, whence $[c_1c_2, a] = [c_2, a][c_1, a]$, and thus it suffices to recall that $\psi_c(a) \equiv \eta([c, a])$.

d) We observe that if $c \in (G_1)_h$ and $a \in H_h$, then by definition

$$\varphi_c(a) = \eta([(s(a))^{-1}, c^{-1}]), \qquad \psi_c(a) = \eta([c, a]).$$

Let us note that $\overline{\varphi_c(a)} = \psi_c(a)$. In fact, we can assume $a \in H_h$. We write $r = s(a)$; then

$$\overline{\varphi_c(a)} = \eta([c^{-1}, r^{-1}]) = \eta([c, r^{-1}]) = \eta([c, r]) = \psi_c(a).$$

Finally, we define $\chi \in \mathrm{Hom}((G_1)_h, \mathrm{X}(K))$ by $\chi(c)(a) = \overline{\psi'_c(a)}$. This depends on the class of $a \in H$ modulo M only. Summing up, we have completed proving the following. There exists $\chi \in \mathrm{Hom}((G_1)_h, \mathrm{X}(K))$, continuous, such that if $c \in (G_1)_h$ then $c\omega' = \iota(\chi(c))\omega'$ in the sense of unitary equivalence.

We proceed now to complete our proof of the Main Lemma. By virtue of what we saw above, we have to prove the following. Assume that $\omega \in \widehat{M}$ is such that $\omega | M^{\mathrm{c}} \not\equiv 1$ and we put $\zeta = \{\omega\} \in \widehat{M}$, and

$$\widehat{H}_\zeta = \{\rho; \rho \in \widehat{H} \text{ such that } \rho | M \approx \omega\}.$$

Then $\widehat{H}_\zeta / G$ is countably separated.

We observe that if it is so, then the G-orbits in $\widehat{H}_\zeta$ are all locally closed and thus, since $\widehat{H}_\zeta$ itself is locally closed in $\widehat{H}$, they are locally closed in $\widehat{H}$ as well. We recall that by virtue of the notation we introduced above, ω' is a $\beta = (\iota(\alpha))^{-1}$ extension of ω from M to H.

(i) We start by noting that we have $G = (G_1)_h.H$. In fact, by what we saw above in Lemma 2, we have $G = G_1.H$ and clearly also $G = (G_1)_h M$. From this we conclude that $G = (G_1)_h MH = (G_1)_h.H$. We note that this implies that it suffices to show that $\widehat{H}_\zeta/(G_1)_h$ is countably separated.

(ii) We write again $K = H/M$ and denote by Λ the set of all unitary equivalence classes of irreducible α-representations of K. For $\lambda \in \Lambda$ we set $\varphi(\lambda) = (\omega' \otimes \iota(\lambda))^{\sim}$.

(iii) Next we introduce a Borel structure on $\widehat{H}_\zeta$ as follows. We denote by a the class of α in $H^2(K, T)$ (cf. [**AM**], p. 25). Let K_a be the corresponding central extension of K by T. We recall that in this situation a representation of K_a is said to be of class one if its restriction to the canonical 1-torus is a multiple of the identity map onto itself. Let $\widehat{K}'_a$ be the set of all irreducible representations that are of class one. We have canonically $\Lambda \sim \widehat{K}'_a$. Evidently $\widehat{K}'_a$ is closed in $\widehat{K}_a$.

Let us put $\mathfrak{A} = C^*(K_a)$. We conclude that there is a closed 2-sided ideal J in $\mathfrak{A}$ such that $\Lambda = (\mathfrak{A}/J)\widehat{}$. We consider Λ in the Borel structure it inherits in this manner. Let us note that Λ is standard. In fact, if not, then $\mathfrak{A}/J$ admits a factor representation that is not of type I (cf. [**D***], 9.5.6, p. 185). If L is such then (cf. [**AM**], pp. 61–62) $(\omega' \otimes \iota(L))^{\sim}$ is a factor representation of the same kind of H. But, by $H = (I)$, this is not possible. We observe next that $\varphi : \Lambda \to \widehat{H}_\zeta$ is a Borel isomorphism. In fact, bearing in mind that $\varphi(\lambda) = (\omega' \otimes \iota(\lambda))^{\sim}$ we see that φ is a Borel injection of Λ onto $\widehat{H}_\zeta$; hence it is enough to observe that 1) Λ and $\widehat{H}_\zeta$ are standard, 2) by the theorem of Souslin (cf. [**AM**], Proposition 2.5, p. 7) φ is a Borel isomorphism. From here we conclude that to arrive at our main goal, it is enough to show that the action of $(G_1)_h$ on $\widehat{K}'_a$, corresponding to that on $\widehat{H}_\zeta$, is countably separated. Below notation will be the same as at the end of 1.2. Thus $K_a = D \times A_a \times A$ and if $\mu \in \widehat{K}'_a$, then $\mu \sim \rho \times \sigma \times \tau$ where ρ and τ are irreducibles on D and A resp. and σ is a class-one irreducible representation on A_a. We denote by χ' the lift of χ to K_a.

We have clearly $(c \in K_a)$

$$\chi'(c)\rho = \rho, \qquad \chi'(c)\sigma = \sigma$$

and $\psi : (G_1)_h \to \widehat{A}$ is a continuous morphism. To see that its orbits are closed, it is enough to recall that $(G_1)_h$ is connected. In fact, we have $G_1 = (G_1)_h M$ and hence $G_1/M = (G_1)_h M/M = (G_1)_h/M_h$, which implies our statement.

1.4. Application to the regular representation of a connected Lie group

The objective of this closing section of Chapter I is the discussion of the regular representation of a connected Lie group. Let G be a separable locally compact group. We denote by dx the element of the right-invariant Haar measure on G, and put $d(sx) = \Delta(s)\, dx$ ($s \in G$, fixed). We can construct on $L^2(G)$ canonically a unitary representation L, called the left-regular representation, as follows. Let f be an element of $L^2(G)$. Then the value $L(s)$ ($s \in G$) is given by $f(x) \mapsto f(s^{-1}x)/\sqrt{\Delta(s)}$. This being so, the left ring of G is the von Neumann algebra, generated by the operators of the left-regular representation; it will be denoted by $L(G)$. If G is unimodular, that is $\Delta = 1$, R. Godement and I. E. Segal proved independently that $L(G)$ is semifinite. This means that $L(G)$ carries a faithful, normal semi-finite trace. In fact, one can write down such Φ, say, as follows. Given $f \in L^1(G) \cap L^2(G)$, let us form the corresponding value in the left ring by

$$L_f = \int_G f(x)L(x)\, dx.$$

Then there is a Φ as above, well-determined by $\Phi(L_f \cdot L_f^*) = \|f\|_2^2$. The original motivation of the question was to construct trace class factor representations, by considering, following Peter and Weyl, the central disintegration into factor representations, of the left-regular representation. In more detail, the said procedure yields a separating family of unitary representations, each of which generates a semifinite factor with the property that, for some choice of $f \in L^1(G) \cap L^2(G)$, its integral with respect to the said representation gives a nonzero trace class operator. We shall take up this question in more detail in Chapter III. Concerning the regular representation, it was pointed out by Godement that the left ring may fail to be semifinite for non-unimodular groups. An example for this is the semi-direct product of $\mathrm{GL}(2, \mathbb{Q})$ with $\mathbb{R}^2$. Our objective in this section will be to prove that for a connected Lie group, $L(G)$ is always semifinite. The proof makes, among others, a substantial use of the Main Lemma.

Let again G be an arbitrary separable locally compact group. The right-regular representation R is defined by the condition $(R(s)f)(x) \equiv f(xs)$, $f \in L^2(G)$, s fixed. The von Neumann algebra $R(G)$, generated by the operators of the right-regular representation, is equal to the commuting ring of the left ring $L(G)$. For $f \in L^2(G)$ let us put $(Sf)(x) \equiv \overline{f(x^{-1})}/\sqrt{\Delta(x)}$. S is an involution, and the map $A \to SAS$ ($A \in L(G)$) takes the left ring on the right ring. Below, given a von Neumann algebra R and a (not necessarily bounded) selfadjoint operator K, we write $K\eta R$

to say that all bounded functions of K belong to R; we shall also say that K is affiliated with R. Given two commuting selfadjoint operators A, B, we shall take the liberty to write just AB for the minimal closed extension of their product. We denote by J the, not necessarily bounded, selfadjoint operator, corresponding to multiplication by the function $\sqrt{\Delta(x)}$. The following proposition, due to Dixmier, of seminal importance for our proof, is the special case of a much more general result.

LEMMA 1. *Unexplained notation as above. Suppose that M' is a positive, selfadjoint, invertible operator, such that $M'\eta L(G), M'$ commutes with J, and if $M = SM'S\eta R(G)$, J is the minimal closed extension of $M' \cdot M^{-1}$. Then $L(G)$ is semifinite.*

For the proof see [**Dq**], Theorem 2, p. 287.

Below, if $A\eta R(G)$, we shall write $A' = SAS\eta L(G)$.

LEMMA 2. *Suppose that M' satisfies all the conditions of Lemma 1, except that $J = M' \cdot M^{-1}$ is replaced by $L(a)M'L(a^{-1}) = (\Delta(a))^{-1/2} \cdot M'$ $(a \in G)$. Then there is an operator C, which is positive, selfadjoint, invertible, $C\eta(L(G))^{\mathrm{c}}$, such that if we define $N = CM$, then J is the minimal closed extension of $N' \cdot N^{-1}$.*

In fact,

(i) We define M_1 by $J = M' \cdot M_1^{-1}$. An easy computation shows that

$$L(a)JL(a^{-1}) = (\Delta(a))^{-1/2} \cdot J.$$

By assumption we have also $L(a)M'L(a^{-1}) = (\Delta(a))^{-1/2} \cdot M'$. We can thus conclude that M_1 is positive, selfadjoint, invertible, affiliated with $R(G)$ and $J = M' \cdot M_1^{-1}$.

(ii) One verifies easily that $SJS = J^{-1}$. We put $M = SM'S$ and $M_1' = SM_1S$; we have respectively $M\eta R(G)$ and $M_1'\eta L(G)$. We have also $J = M_1' \cdot M^{-1}$. In fact,

$$\begin{aligned} J^{-1} &= SJS = S(M' \cdot M_1^{-1})S \\ &= SM'S \cdot (SM_1S)^{-1} = M \cdot M_1'^{-1}; \text{ hence } J = M_1' \cdot M^{-1}. \end{aligned}$$

Summing up, we have so far $J = M' \cdot M_1^{-1}$ (by definition) and $J = M_1' \cdot M^{-1}$.

(iii) We observe next that M and M_1 commute. In fact, we have by definition $M_1 = J^{-1} \cdot M'$. Then M commutes with M' and, by $M = SM'S$ and $J^{-1} = SJS$, M commutes with J^{-1}.

Conclusion. So far $J = M' \cdot M_1^{-1}$, $J = M_1' \cdot M^{-1}$ and M commutes with M_1, hence M' commutes with M_1'.

(iv) We conclude from (iii) that $M' \cdot M_1^{-1} = M_1' \cdot M^{-1}$ whence $(M_1')^{-1} \cdot M' = M^{-1} \cdot M_1$, and that there is a C positive, selfadjoint, invertible, $C\eta(L(G))^{\mathrm{c}}$ such that $M \cdot M_1^{-1} = C$, $(M')^{-1} \cdot M_1' = C$, and $SCS = C^{-1}$.

(v) We have

$$J = M' \cdot M_1^{-1} = M' \cdot M^{-1} \cdot MM_1^{-1} = CM' \cdot M^{-1}.$$

(vi) We define $D > 0$ by $C = D^2$; then also $SDS = D^{-1}$ and $J = D \cdot M' \cdot DM^{-1}$. We write $N = D^{-1}M$; this implies

$$N' = SD^{-1} \cdot MS = S(D^{-1})S \cdot M = D \cdot M'$$

and thus $J = N' \cdot N^{-1}$; hence it suffices to take C in place of D^{-1}.

LEMMA 3. *Unexplained notation as above. Suppose that G is a connected Lie group, and that H is a closed, unimodular invariant subgroup of G, which is of type I, such that G/H is abelian. Let μ be the Plancherel measure on $\widehat{H}$. Suppose that there is a positive, μ-measurable function m on $\widehat{H}$, such that for any a fixed in G, we have for almost all $\zeta : m(a\zeta) \equiv \Delta(a)m(\zeta)$. Then $L(G)$ is semi-finite.*

In fact, by virtue of Lemmas 1 and 2 it is enough to show that there is an operator M' on $L^2(G)$, which is positive, selfadjoint, invertible, is affiliated with $L(G)$, commutes with J and satisfies $L(a)M'L(a^{-1}) = (\Delta(a))^{-1/2}M'$ for all $a \in G$.

(i) We recall (cf. [**AM**], Proposition 3.2, p. 14), that there is a Borel map f from G/H into G, such that $f(y) \in y$ for all y in G/H. Since G/H is standard, by virtue of the theorem of Souslin (cf. [**AM**], Proposition 2.5, p. 7) $S = f(G/H)$ is a Borel subset of G and we have $G = SH$ setwise; finally, f is a Borel isomorphism from G/H onto S.

(ii) We have, under the condition of proper normalization of Haar measures dh and dv on H and G/H respectively, for any function f with compact support on G:

$$\int_G f(x) \cdot dx = \int_{G/H} \left(\int_H f(hy)\, dh \right) dv.$$

Let $d\tau$ be the image of dv on S. We denote by L_H the left regular representation on $L^2(H)$. Then there is a canonical identification between $L^2(G)$ and $L^2(H) \otimes L^2_\tau(S)$, which makes correspond $L_H \otimes 1$ to $L|H$. We observe that evidently, for any $a \in G$, there is a *-automorphism ψ_a of $L(H)$, such that $\psi_a(L_H(h)) = L_H(aha^{-1})$ $(h \in H)$. Its domain of definition may be extended to all positive selfadjoint operators, by agreeing that, if

$$A = \int_0^{+\infty} \lambda \cdot dE_\lambda$$

then

$$\psi_a(A) = \int_0^{+\infty} \lambda \cdot d\psi_a(E_\lambda).$$

(iii) Let K be a positive, selfadjoint, invertible operator on $L^2(H)$ such that $K\eta L(H)$ and $\psi_a(K) = (\Delta(a))^{-1/2} \cdot K$ for all a in G. We put $M' = (K \otimes I)\eta L(G)$ and claim that M' satisfies all the conditions of Lemma 2. In fact, to prove our point it suffices to establish the following two statements:

a) M' commutes with J,

b) $L(a)M'L(a^{-1}) = (\Delta(a))^{-1/2} \cdot M'$ $(a \in G)$. In fact,

a) We observe that the image of J on $L^2(H) \otimes L^2(S)$ is multiplication with the function $(\Delta(hs))^{1/2}$ $(h \in H, s \in S)$. Hence it is enough to show $\Delta(h) \equiv 1$ $(h \in H)$. This, however, is implied by the condition that H be unimodular, and G/H abelian.

b) Here we have:

$$\begin{aligned} L(a)M'L(a^{-1}) &= L(a)(K \otimes I)L(a^{-1}) \\ &= \psi_a(K) \otimes I = (\Delta(a))^{-1/2}(K \otimes I) = (\Delta(a))^{-1/2} \cdot M'. \end{aligned}$$

Summing up, to complete the proof of our lemma, it suffices to construct the operator K as just considered.

(iv) In the following we continue to assume that H is unimodular and of type I. Before proceeding, we recall a few things concerning the Plancherel formula of such groups (cf. [**D***], pp. 327–328). Let $\zeta \mapsto \pi(\zeta)$ ($\zeta \in \widehat{H}$) be a field, measurable for all positive measure on $\widehat{H}$, of irreducible representations such that $\pi(\zeta)$ is of the class $\zeta \in \widehat{H}$. There is a positive measure μ on $\widehat{H}$ (called the Plancherel measure on $\widehat{H}$), well-determined by the following.

If $u \in L^1(H) \cap L^2(H)$ and $\zeta(u) = \pi(\zeta)(u)$ ($\zeta \in \widehat{H}$) then

$$\int_H |u(h)|^2 \cdot dh = \int_{\widehat{H}} \operatorname{Tr}(\zeta(u)(\zeta(u))^*) \cdot d\mu(\zeta).$$

The correspondence $u \mapsto \pi(\zeta)(u)$ extends to a unitary isomorphism from $L^2(H)$ onto the set of all measurable fields of Hilbert-Schmidt operators $(A(\zeta))$ such that

$$\int_{\widehat{H}} \operatorname{Tr}(A(\zeta)(A(\zeta))^*) \cdot d\mu(\zeta) < +\infty.$$

We shall write $\mathfrak{H}$ for this space. This correspondence, in particular, sends $(L(H))^{\mathrm{c}}$ onto the set of all multiplications by elements of $L^\infty_\mu(\widehat{H})$. Given $\varphi \in L^\infty_\mu(\widehat{H})$ we write $M(\varphi)$ for the corresponding multiplication operator. Our next objective is to determine the action of ψ_a on $L^\infty_\mu(\widehat{H})$ ($\sim (L(H))^{\mathrm{c}}$). For $a \in G$ and $\varphi \in L^\infty_\mu(\widehat{H})$ we write $\varphi^a(\zeta) \equiv \varphi(a^{-1}\zeta)$. This being so we claim that $\psi_a(M(\varphi)) = M(\varphi^a)$. In fact

a) Since G/H is abelian, we have for all $a \in G$: $d(aha^{-1}) = \Delta(a) \cdot dh$.

b) 1) We define the unitary operator V on $L^2(H)$ by $(Vf)(h) = (\Delta(a))^{-1/2} \cdot f(a^{-1}ha)$. Evidently, if $T \in L(H)$, then $\psi_a(T) \equiv VTV^{-1}$ ($a \in G$, fixed).

2) If $f \in L^1(H) \cap L^2(H)$, then its image in $\mathfrak{H}$, A say, satisfies $A(\zeta) \equiv \pi(\zeta)(f)$. Hence if W corresponds to V on $\mathfrak{H}$, then $W\{\pi(\zeta)(f)\} = \{\pi(\zeta)V(f)\}$. Let A be as above and $B = W(A)$. We claim that

$$B(\zeta) \equiv (\Delta(a))^{1/2} \cdot UA(a^{-1}\zeta)U^*$$

where U depends on ζ only. In fact

$$\begin{aligned} B(\zeta) &= \int_H \pi(\zeta)(h) \cdot (\Delta(a))^{-1/2} \cdot f(a^{-1}ha) \cdot dh \\ &= (\Delta(a))^{1/2} \cdot \int_H a^{-1}[\pi(\zeta)](h) \cdot f(h) \cdot dh. \end{aligned}$$

We note that there is a unitary operator U such that $a^{-1}\pi(\zeta) = U\pi(a^{-1}\zeta) \cdot U^*$, whence

$$B(\zeta) = (\Delta(a))^{1/2} \cdot UA(a^{-1}\zeta)U^*.$$

3) From this it follows at once that $WM(\varphi)W^{-1} = M(\varphi^a)$, as claimed above.

(v) We recall (cf. (iii)) that our goal is to construct an operator K on $L^2(H)$, which is positive, selfadjoint, invertible, is affiliated with $L(H)$ and satisfies

$$\psi_a(K) = (\Delta(a))^{1/2}K \quad (a \in G).$$

We are given a positive, μ-measurable function on $\widehat{H}$, such that for each $a \in G$, $m(a\zeta) = \Delta(a)m(\zeta)$ almost everywhere in ζ with respect to μ. Hence, by (iv), the operator on $L^2(H)$, corresponding to $M((m)^{1/2})$ will satisfy all the conditions.

Our next objective is the construction of the group H and of the function m appearing in the statement of Lemma 3.

LEMMA 4. *Let G be a connected Lie group with the Lie algebra $\mathfrak{g}$. If $\mathfrak{m}$ is the greatest nilpotent ideal in $\mathfrak{g}$, we set $\mathfrak{h} = [\mathfrak{g}, \mathfrak{g}] + \mathfrak{m}$. Let H be the connected subgroup corresponding to $\mathfrak{h}$ of G. Then it is closed in G.*

The proof will be based on the following

SUBLEMMA. *Let G be a simply connected group with the Lie algebra $\mathfrak{g}$. Assume $\mathfrak{h}$ to be defined as in Lemma 4. We form the corresponding connected subgroup H (which is now closed). Let $\varphi : G \to G/H$ be the canonical morphism. Then, if Z is the center of G, $\varphi(Z)$ is discrete in G/H.*

In fact,

(i) We observe that if $\mathfrak{i}$ is a Levi section in $\mathfrak{g}$, then $\mathfrak{h} = \mathfrak{i} + \mathfrak{m}$.

In fact,

a) We note first that $\mathfrak{h} \subseteqq \mathfrak{i} + \mathfrak{m}$. Since $\mathfrak{h} = [\mathfrak{g}, \mathfrak{g}] + \mathfrak{m}$, it suffices to show that $[\mathfrak{g}, \mathfrak{g}] \subseteqq \mathfrak{i} + \mathfrak{m}$. Let $\mathfrak{u}$ be the radical of $\mathfrak{g}$; then we have $\mathfrak{g} = \mathfrak{i} + \mathfrak{u}$ (direct sum of subspaces). We have also: $[\mathfrak{g}, \mathfrak{g}] \subseteqq [\mathfrak{i}, \mathfrak{i}] + [\mathfrak{i}, \mathfrak{u}] + [\mathfrak{u}, \mathfrak{u}]$. Since $\mathfrak{i}$ is semi-simple, $[\mathfrak{i}, \mathfrak{i}]$ is equal to $\mathfrak{i}$. Next we recall (cf. [**B**], Proposition 6, p. 79) that if D is a derivation of $\mathfrak{g}$, then $D\mathfrak{u} \subseteqq \mathfrak{m}$. From this we conclude that $[\mathfrak{i}, \mathfrak{u}] \subseteqq \mathfrak{m}$, $[\mathfrak{u}, \mathfrak{u}] \subseteq \mathfrak{m}$, completing the proof of $\mathfrak{h} \subseteqq \mathfrak{i} + \mathfrak{m}$.

b) We show next that $\mathfrak{i} + \mathfrak{m} \subseteqq \mathfrak{h}$. In fact, this follows at once from $\mathfrak{i} = [\mathfrak{i}, \mathfrak{i}] \subseteqq [\mathfrak{g}, \mathfrak{g}]$.

Summing up, we have shown that if $\mathfrak{i}$ is a Levi section, then $\mathfrak{h} = \mathfrak{i} + \mathfrak{m}$.

(ii) Next we show that $\mathrm{Ad}(H)$ is closed in $\mathrm{GL}(\mathfrak{g})$. To this end it suffices to prove that $\mathrm{ad}(\mathfrak{h})$ is algebraic in $\mathrm{GL}(\mathfrak{g})$. Since $\ker(\mathrm{ad}(\mathfrak{i}))$ is zero, $\mathrm{ad}(\mathfrak{i})$ is semi-simple and thus algebraic. The endomorphisms in $\mathrm{ad}(\mathfrak{m})$ are all nilpotent; hence $\mathrm{ad}(\mathfrak{m})$ is algebraic. Since $\mathrm{ad}(\mathfrak{h}) = \mathrm{ad}(\mathfrak{i}) + \mathrm{ad}(\mathfrak{m})$, $\mathrm{ad}(\mathfrak{h})$ itself is algebraic (cf. [**Ch II**], Th. 14, p. 175). Hence $\mathrm{Ad}(H)$ is closed in $\mathrm{GL}(\mathfrak{g})$.

(iii) We recall that Z is the center of G. Since ZH is the complete inverse image through Ad of $\mathrm{Ad}(H)$, by (ii) it is closed. If φ is the canonical morphism $\varphi : G \to G/H$, then $\varphi(Z)$ is closed in G/H. Finally, since $Z_0 \subseteqq H$, $\varphi(Z)$ is discrete in G/H, completing the proof of the Sublemma.

We now turn to the demonstration of Lemma 4. Let G be as given loc. cit. We write (G_1, π) for the universal covering of G, and set $K = \ker(\pi)$. Let Z_1 be the center of G_1 and H_1 the subgroup corresponding to $\mathfrak{h}$ of G_1. We have $H = \pi(H_1)$ and thus

$$\pi^{-1}(H) = \pi^{-1}(\pi(H)) = H_1 \cdot K.$$

Let us write ψ for the canonical morphism $G_1 \to G_1/H_1$. The Sublemma implies that $\psi(Z_1)$ is discrete in G_1/H_1. Hence $\psi(K)$ is discrete and thus also closed in G_1/H_1. In this manner $H_1 \cdot K = \psi^{-1}(\psi(K))$ is closed in G_1. Therefore $H = \pi(H_1) = \pi(H_1 \cdot K)$ is closed in G, completing the proof of Lemma 4.

LEMMA 5. *With the above notations, the group H is unimodular, of type* I *and such that G/H is abelian.*

In fact, since evidently $\mathrm{Tr}(\mathrm{ad}\, l) \equiv 0$ ($l \in \mathfrak{h}$), H is unimodular. By 1.2 it is of type I, since $[\mathfrak{g}, \mathfrak{g}] + \mathfrak{m}$ is isomorphic to an algebraic Lie algebra. Finally, G/H is abelian by $\mathfrak{h} \supseteqq [\mathfrak{g}, \mathfrak{g}]$.

The following Lemma 6 is literally the same as Lemma 3.1 in [**Dreg**], p. 427.

LEMMA 6. *Let K be a separable, locally compact abelian group. We assume that S is a standard left K-space such that S/K is countably separated. Let μ*

be a positive, σ-finite measure on S. Suppose that χ is a continuous morphism from K into the multiplicative group of positive reals such that for any $k \in K$ we have $k\mu = \chi(k)\mu$. Then there is a positive, measurable function $m : S \to \mathbb{R}$ such that $m(s) > 0$ almost everywhere with respect to μ and such that for each k fixed, $m(ks) = \chi(k)m(s)$.

To prove this, let us write $\pi : S \to S/K$ for the canonical projection.

(i) We start by noting that S/K is analytic. In fact, this follows from the following statement, which we have already used on several occasions (cf. [**AM**], Proposition 2.11, p. 9, in particular). Assume that E is analytic, F countably separated and $f : E \to F$ is Borel and surjective. Then E/f is analytic. We apply this in our present situation by taking $E = S$ (standard by hypothesis), $F = S/K$ (countably separated by hypothesis), and $\pi(x) \equiv Kx$ ($x \in S$). We conclude from all this, that S/K is analytic.

(ii) Let ν be a measure, which is finite and equivalent to μ on S; we put $\nu' = \pi_*(\nu)$. We recall (cf. [**AM**], p. 11, bottom) that if E is analytic and τ a σ-finite measure, then there is a standard subspace F of E such that $\tau(E-F) = 0$. We conclude from this that we can assume S/K to be standard.

(iii) We next recall the following result (cf. [**AM**], Proposition 2.15, p. 12). Assume that E is standard and F analytic. We suppose that $p : E \to F$ is Borel onto. Let λ be a σ-finite measure on F. There exists a Borel set Σ in F such that $\lambda(F - \Sigma) = 0$, a Borel map $\varphi : \Sigma \to p^{-1}(\Sigma)$ such that $p \circ \varphi = \mathrm{id}_\Sigma$. We conclude from this that S/K can be permitted to be standard, and φ defined on the whole of S/K such that $\pi \circ \varphi = \mathrm{id}_{S/K}$.

(iv) We observe next that $\varphi(S/K)$ is Borel in S. In fact, to see this, it is enough to recall again (cf. [**AM**], Proposition 2.5, p. 7, in particular) that if E, F are standard, $f : E \to F$ is injective, Borel, then $f(E)$ is Borel in F. Conclusion: $\varphi(S/K)$ ($= T$, say) is Borel in S.

(v) (cf. (ii) above) We form $\psi : K \times T \to S$ such that $\psi(k,t) = kt$. Then S is Borel isomorphic to $(K \times T)/\psi$. Hence we can assume $S = (K \times T)/\psi$ such that $(k,t) \sim (k',t')$ if and only if $t = t'$ and $kt = k't'$, which implies $k' = kk_0$ ($k_0 t = t$).

(vi) Before proceeding, we recall the following facts.

A) (cf. [**Dreg**], p. 428).

There is on S a Polish locally compact structure such that 1) The Borel structure resulting from this topology coincides with the given one; 2) μ is a Radon measure.

B) (cf. [**B int**], pp. 64–68).

The equivalence relation defined by K on S is μ-measurable. From this we are permitted to infer that for each $y \in S/K$ there is a positive, σ-finite measure on $\pi^{-1}(y)$, such that for each positive, Borel measurable function f we have

$$\int_S f(s) \cdot d\mu(s) = \int_{S/K} d\nu'(y) \left(\int_{\pi^{-1}(y)} f(t) \cdot d\mu_y(t) \right).$$

If $(\mu'_y)_{y \in S/K}$ is another family of the said sort, then $\mu_y = \mu'_y$ almost everywhere with respect to φ'.

Assume that $k \in K$ is fixed. Then, by virtue of our starting assumption, we have: $k(\mu_y) = \chi(k)\mu_y$ for almost all y with respect to ν', the set of exceptions being permitted to depend on $k \in K$. In fact, let f be a positive Borel function on S;

then we have

$$\begin{aligned}\chi(k)\int_S f(s)\cdot d\mu(s) &= (\text{by } k\mu = \chi(k)\mu)\\ &= \int_S f(s)\cdot d(k\mu)(s) = \int_S f(ks)\cdot d\mu(s)\\ &= \int_{S/K} d\nu'(y)\int_{\pi^{-1}(y)} f(kt)\cdot d\mu_y(t)\\ &= \int_{S/K} d\nu'(y)\int_{\pi^{-1}(y)} f(t)\cdot d(k\mu_y)(t)\end{aligned}$$

whence $k\mu_y = \chi(b)\mu_y$ for almost all y with respect to ν'.

(vii) We conclude from this that there is a sequence $(k_1, k_2, \dots)$ dense in K, there is a subset N of S/K of ν'-measure zero, such that if y does not belong to N, then we have $k(\mu_y) = \chi(k)\mu_y$ for any $k = k_i$ $(i = 1, 2, \dots)$. Since $k \mapsto kt$ ($k \in K$; t is fixed such that $\pi(t) = y$) is Borel, there is on $\pi^{-1}(y)$ a structure of a K-homogeneous space, giving rise to our Borel structure.

For reasons of continuity, $k\mu_y = \chi(k)\mu_y$ $(y \in N)$ for all k in K.

(viii) We claim that there is $m > 0$, measurable on S such that $m(ks) = \chi(k)m(s)$ for each $k \in K$ and almost all s in S. In fact,

a) Assume that $\psi((k_1, t_1)) = \psi((k_2, t_2))$. Then $\chi(k_1) = \chi(k_2)$. In fact, $\psi((k_1, t_1)) = \psi((k_2, t_2))$ is equivalent to $t_1 = t_2$ ($= t$, say) and $k_1t_1 = k_2t_2$. Putting $k = k_2^{-1}\cdot k_1$ we have $t = kt$, whence, since K is abelian, $ku \equiv u$ $(u \in kt)$ and $k\mu_y = \mu_y$ $(y \in K)$ giving $\chi(k) = 1$ and also $\chi(k_1) = \chi(k_2)$ (if $\psi((k_1, t_1)) = \psi((k_2, t_2))$). We define $m(s) = \chi(k)$ if $s = (k, t)$.

b) If $s \in S - \pi^{-1}(N)$ and $k \in K$, then $m(ks) = \chi(k)m(s)$. In fact, assume $s = k't$. Then we have:

$$m(ks) = m(kk't) = m(k)\chi(k') = \chi(k)m(s)$$

yielding $m(ks) = \chi(k)m(s)$. This completes the proof of Lemma 6.

Editors' note. For another proof of this lemma, see A. Brunel, Nghiem Dang-Ngoc, and J.-P. Thouvenot, *Un problème d'existence de mesures invariantes*, Ann. Inst. H. Poincaré, Ser. B (N.S.) **10**, 1974, pp. 211–227 (especially pp. 212–213).

LEMMA 7. *We assume that G is a separable, locally compact group, with the right-invariant Haar measure dx. For $a \in G$ we set $d(ax) = \Delta(a)\,dx$. We assume that H is a closed, invariant, unimodular, type* I *subgroup, such that G/H is abelian. We write μ for the Plancherel measure of H. Then, for all $a \in G$ we have*

$$d(aha^{-1}) = (\Delta(a))^{-1}\cdot dh \quad \textit{and} \quad d\mu(a\zeta) = \Delta(a)\cdot d\mu(\zeta).$$

In fact,

(i) We note first that $d(aha^{-1}) = \Delta(a)dh$. To see this, we observe that

$$\int_G f(aga^{-1})\cdot dg = \int_G f(g)d(a^{-1}ga) = (1/\Delta(a))\int_G f(g)\cdot dg.$$

On the other hand, if on G/H, dv is appropriately chosen, then

$$\int_G f(g)\cdot dg = \int_{G/H}\left(\int_H f(xh)\cdot dh\right)dv$$

and thus

$$\int_G f(aga^{-1}) \cdot dg = \int_{G/H} \left(\int_H f(axha^{-1}) \cdot dh \right) dv$$
$$= \int_{G/H} \left(\int_H f(axa^{-1} \cdot aha^{-1}) \cdot dh \right) dv.$$

We write $d(a^{-1}ha) = k(a) \cdot dh$; hence

$$\int_G f(aga^{-1}) \cdot dg = k(a) \int_{G/H} \left(\int_H f(rh) \cdot dh \right) dv = k(a) \int_G f(g) \cdot dg.$$

We conclude from this that $d(aka^{-1}) = (\Delta(a))^{-1} \cdot dh$.

(ii) We complete now the proof that $d\mu(a\zeta) = \Delta(a) \cdot d\mu(\zeta)$ in the following manner.

We have

$$\int_H |f(h)|^2 \cdot dh = \int_{\widehat{H}} [\pi(\zeta)(f)]^2 \cdot d\mu(\zeta).$$

Let us put $f_a(h) \equiv f(aha^{-1})$. Then

$$\int_H |f_a(h)|^2 \cdot dh = \Delta(a) \int_H |f(h)|^2 \cdot dh.$$

We have

$$\pi(\zeta)(f_a) = \int_H \pi(\zeta)(h) f(aha^{-1}) \cdot dh$$
$$= \Delta(a) \cdot \int_H \pi(\zeta)(a^{-1}ha) \cdot f(h)\, dh = \Delta(a) \cdot W \left(\int_H \pi(a\zeta) f(h) \cdot dh \right) W^*,$$

where W is a constant unitary, or

$$\pi(\zeta)(f_a) = \Delta(a) \cdot w\pi(a\zeta) \cdot W^*$$

and thus

$$\|f_a\|^2 = (\Delta(a))^2 \int_{\widehat{H}} [\pi(a\zeta)(f)]^2 \cdot d\mu(\zeta)$$
$$= (\Delta(a))^2 \cdot \int_{\widehat{H}} [\pi(\zeta)(f)]^2 \cdot d\mu(a^{-1}\zeta).$$

On the other hand,

$$\|f_a\|^2 = \Delta(a) \cdot \int_H |f(h)|^2 \cdot dh = \Delta(a) \int_{\widehat{H}} [\pi(\zeta)(f)]^2 \, d\mu(\zeta),$$

whence the conclusion that $d\mu(a\zeta) = \Delta(a) \cdot d\mu(\zeta)$.

Now we are ready to prove the

THEOREM. *Suppose that G is a connected Lie group. Then the left (right) ring is semi-finite.*

Let $\mathfrak{g}$ be the Lie algebra of G.

(i) As in Lemma 4, we set $\mathfrak{h} = [\mathfrak{g}, \mathfrak{g}] + \mathfrak{m}$. We recall that if H is the connected subgroup of G corresponding to $\mathfrak{h}$, then H is closed.

(ii) By virtue of the Theorem of Ado (cf. [**Ch III**], p. 153) we can assume that $\mathfrak{g}$ is isomorphic to some subalgebra of the Lie algebra of all endomorphisms of a finite-dimensional real vector space V in such a manner that the greatest nilpotent ideal of $\mathfrak{g}$ will be composed of nilpotent endomorphisms. We recall (cf. [**Ch II**], p. 173) that $[\mathfrak{g}, \mathfrak{g}]$ is always algebraic; hence $\mathfrak{h}$, too, is algebraic.

(iii) Let $\mathfrak{a}$ be the smallest algebraic Lie algebra containing $\mathfrak{g}$. We have $[\mathfrak{a}, \mathfrak{a}] = [\mathfrak{g}, \mathfrak{g}]$ (cf. loc. cit.).

Let A and H_1 be the irreducible subgroups of $\mathrm{GL}(V)$ corresponding to $\mathfrak{a}$ and $\mathfrak{h}$ respectively. Then H_1 is invariant in A.

(iv) We put $H_2 = (H_1)_0$; this is a connected Lie group, closed in $\mathrm{GL}(V)$. Let (H_3, η) be the universal covering of H_2.

Since $\mathrm{Lie}(H_2) = \mathfrak{h} = \mathrm{Lie}(H)$, there is a homomorphism from H_3 onto H, φ say, the kernel T of which is discrete. Evidently, $\eta(T)$ is the center of H_2.

(v) Let a be some element in A. We define $\rho_2(a)h_2 = ah_2a^{-1}$ $(h_2 \in H_2)$. Let $\rho_3(a)$ be the covering map of $\rho_2(a)$ $(a \in H_2)$. We form

$$D' = \{b; b \in A \text{ such that } \rho_2(b)\eta(t) \equiv \eta(t) \text{ for all } t \in T\}.$$

We note that 1) D' is algebraic in $\mathrm{GL}(V)$ and 2) $H_2 \subset D'$. We define $D = (D')_0$.

(vi)

1) Assume that $b \in D$. Then there is $\tau(b) \in \mathrm{Aut}(H)$ such that $(\tau(b))_* = \mathrm{Ad}(b; \mathfrak{h})$. (Note that, evidently, $\tau(b)$ is well-determined.) In fact, if $b \in D'$ we have, by construction, $\rho_2(b)\eta(t) \equiv \eta(t)$ $(t \in T)$; hence $\rho_3(b)t \in t \cdot \ker(\eta)$. If $b \in D$ $(= (D')_0)$, then $\rho_3(b)t \equiv t$ $(t \in T)$. By $H = H_3/T$ we conclude from this that there is $\tau(b) \in \mathrm{Aut}(H)$, such that $(\tau(b))_* = \mathrm{Ad}(b; \mathfrak{h})$. We define for $g \in G : \sigma(g) \in \mathrm{Aut}(H)$ such that $\sigma(g)h = ghg^{-1}$.

2) We claim that $\sigma(G) \subset \tau(G)$. In fact, we have for $x \in \mathfrak{g}, \lambda \in \mathbb{R}$

$$\mathrm{Ad}(\exp_A(\lambda x))|\mathfrak{h} = \exp(\lambda\, \mathrm{ad}(x; \mathfrak{h})) = \mathrm{Ad}(\exp_G(\lambda x))|\mathfrak{h}.$$

Therefore $\rho_3(\exp_A(\lambda x))$, by passage to the quotient, defines the automorphism $\sigma(\exp_G(\lambda x))$ of H. Therefore, $\exp_A(\lambda x) \in D'$ and also $\sigma(\exp_G(x)) \in \tau(D)$ and $\sigma(G) \subset \tau(G)$ since G is connected.

(vii) We recall that $D \supseteqq H_2$. By the Main Lemma (cf. 1.3) the space $\widehat{H}_3/D$ is countably separated. Since by $H = H_3/T$, $\widehat{H}$ is a closed, D-invariant subset of $\widehat{H}_3$, the same is valid for $\widehat{H}/D$.

(viii) We write D_1 for the commutator group of D.

1) We claim that D_1 operates trivially in $\widehat{H}$. In fact, to show this, it is enough to observe that by $[\mathfrak{a}, \mathfrak{a}] = [\mathfrak{g}, \mathfrak{g}] \subset \mathfrak{h}$, any element of D_1 operates on $\widehat{H}$ as $\tau(h)$ for some $h \in H$.

2) We claim that $\overline{D}_1$, too, acts trivially on $\widehat{H}$. This follows from 1) jointly with the fact (cf. [**AM**], Proposition 3.7, p. 16) that if E is a countably separated Borel G-space, where G is locally compact, then G_x is closed for all $x \in E$.

(ix) We set $K = D/\overline{D}_1$. Then K is locally compact, abelian, with a countably separated action on $\widehat{H}$ (cf. (vii)). There is a positive, continuous character χ on K such that if ν is the Haar measure on H, then $\tau(b)\nu = \chi(b)\nu$ and thus, if μ is the Plancherel measure on $\widehat{H}$, then $\tau(b)\mu = (\chi(b))^{-1}\mu$ (cf. Lemma 7). By Lemma 6, there is a positive Borel function m on $\widehat{H}$ such that for any fixed $b \in D$, $m(bs) = \chi(b)m(s)$ almost everywhere with respect to s in $\widehat{H}$.

Suppose that g in G. Then, by (vi), 2) there is a $b \in D$ such that $\sigma(g) = \tau(b)$. We have then

$$\chi(b)\nu = \tau(b)\nu = \sigma(g)\nu = (\Delta(g))^{-1}\nu \quad \text{(cf. Lemma 7)}.$$

We conclude from this that $\chi(b) = (\Delta(g))^{-1}$ and

$$m(gs) = (\Delta(g))^{-1} \cdot m(s).$$

Thus it suffices to take $1/m$ in place of m.

CHAPTER II

Representations of Elementary Groups

We say that a group Γ is elementary if it is a central extension by a 1-torus of a closed subgroup of $\mathbb{R}^n$. In this chapter our main object of inquiry will be some aspects of the representation theory of such groups; these results will be used substantially in the following two chapters. The necessity of the questions to be considered below should not be surprising, since in the following chapter we propose to analyze the representation theory of a connected Lie group in terms of that of a connected subgroup, which gives rise to an abelian quotient. In this case the Mackey group will be an elementary group in the sense of the above definition. The principal results are easy to state. They fall into two sets.

Let U be the centralizer of the connected center of Γ. We denote by $\mathcal{F}$ the set of all those characters of U^{c} which, when restricted to the canonical torus of Γ, give rise to the identity map of the latter onto itself. Let χ be an element of $\mathcal{F}$. Then the representation induced by χ in Γ, $V(\chi)$ say, is a semifinite factor representation of Γ. It is of type one if and only if the subgroup $U^{\mathrm{c}}\Gamma_0$ is of finite index in U.

Finally, for χ and χ' in $\mathcal{F}$, $V(\chi)$ and $V(\chi')$ are quasi-equivalent if and only if χ and χ' lie on the same orbit of Γ on $\mathcal{F}$; in this case they are also unitarily equivalent (cf. for this 2.2, Proposition).

Next we observe that, given a factor representation T of Γ, its restriction to the canonical torus is a multiple of a character of the latter. We shall say that T is of class one if this restriction is a multiple of the identity map. Clearly, if T_1 is of class one, and $\ker(T_1) = \ker(T_2)$, then T_2, too, is of class one. Similar observations apply in connection with representations of class minus one.

Next we recall (cf. [**Dal**], p. 100) that $\ker(T)$ is a primitive ideal of $C^*(\Gamma)$. In this manner we can speak of a primitive ideal of class one (or minus one). Let us denote by $\mathrm{Prim}_1(\Gamma)$ the set of all primitive ideals of class one of Γ.

If $\chi \in \mathcal{F}$, we put $J(\chi) = \ker(V(\chi))$; this is an element of $\mathrm{Prim}_1(\Gamma)$. This being so we shall show (cf. Proposition in 2.8) that $\chi \mapsto J(\chi)$ is a surjective map from $\mathcal{F}$ onto $\mathrm{Prim}_1(\Gamma)$. We have $J(\chi_1) = J(\chi_2)$ if and only if the closures of the Γ orbits of χ_1 and χ_2 in $\mathcal{F}$ coincide.

2.1. Special case: Extensions of free abelian groups

In this first section we shall discuss the special case of the Proposition of 2.2, when Γ is a central extension by a 1-torus of a free abelian group (such a group will always be understood to be finitely generated).

PROPOSITION. *Suppose that the group Γ satisfies*

$$1 \to \mathcal{T} \to \Gamma \to \mathcal{Z} \to 1$$

where $\mathcal{T}$ is a 1-torus and $\mathcal{Z}$ is a free abelian group. Let χ be a character of Γ^{c} which, when restricted to $\mathcal{T}$, gives rise to the identity map of the latter onto itself.

We put

$$\operatorname{ind}_{\Gamma^c\uparrow\Gamma}\chi = V(\chi).$$

Then $V(\chi)$ is a factor representation of finite type and of class one of Γ. It is of type one if and only if Γ^c is cofinite in Γ.

We start by recalling (cf. [**AM**], p. 188) that there is a bilinear skew-symmetric map $\alpha : \mathcal{Z}\times\mathcal{Z}\to\mathcal{T}$ such that

$$\Gamma = \{(z,u); z\in\mathcal{Z}, u\in\mathcal{T}\}$$

with the law of composition $(z,u)(z',u') = (z+z', \alpha(z,z')uu')$. Given a subset $\mathcal{B}$ of $\mathcal{Z}$, we shall write B for $\{(b,u); b\in\mathcal{B}, u\in\mathcal{T}\}$. Hence, in particular, $\Gamma = Z$.

1) We put

$$\mathcal{Z}_0 = \{x\in\mathcal{Z} \text{ such that } (\alpha(x,y))^2\equiv 1 \text{ for all } y\in\mathcal{Z}\}.$$

Clearly $Z_0 = \Gamma^c$.

2) We write

$$\mathcal{Z}_1 = \{x\in\mathcal{Z} \text{ such that } \alpha(x,y)\equiv 1 \text{ for all } y\in\mathcal{Z}\}.$$

We have $\mathcal{Z}_1\subset\mathcal{Z}_0$.

We write $\chi_1 = \chi|Z_1$ and

$$\mathcal{G} = \{\varphi; \varphi\in \mathrm{X}(Z_0) \text{ such that } \varphi|Z_1\equiv\chi_1\}.$$

LEMMA 1. *We put $V = \operatorname{ind}_{Z_1\uparrow Z}\chi_1$ and, for $\varphi\in\mathcal{G}, U_\varphi = \operatorname{ind}_{Z_0\uparrow Z}\varphi$. We claim that*

$$V = \bigoplus_{\varphi\in\mathcal{G}} U_\varphi$$

and $U_\varphi|Z_0$ is a multiple of φ.

a) We note that Z_0/Z_1 is finite. In fact, this is a finitely generated abelian group, where the order of each element is ≤ 2.

b) We observe that

$$\begin{aligned} V &= \operatorname{ind}_{Z_1\uparrow\Gamma}\chi_1 = \operatorname{ind}_{Z_0\uparrow\Gamma}(\operatorname{ind}_{Z_1\uparrow Z_0}\chi_1) \\ &= \bigoplus_{\varphi\in\mathcal{G}}\Big(\operatorname{ind}_{Z_0\uparrow\Gamma}\varphi\Big) = \bigoplus_{\varphi\in\mathcal{G}} U_\varphi. \end{aligned}$$

Since $Z_0 = \Gamma^c$, $U_\varphi|Z_0$ is a multiple of φ.

We assume that $\mathcal{A}$ is a countable abelian group. Let $\beta : \mathcal{A}\times\mathcal{A}\to\mathbb{T}$ be bilinear and skew-symmetric. We consider the set

$$\overline{\mathcal{A}} = \{(a,u); a\in\mathcal{A}, u\in\mathbb{T}\}$$

and define on it the structure of a group by

$$(a,u)(b,v) = (a+b, \beta(a,b)uv).$$

We define $\chi_0\in\mathrm{X}(\mathbb{T})$ by $\chi_0(0,u)\equiv u$ $(u\in\mathbb{T})$. Given a Hilbert space H, below we write $B(H)$ for its full ring.

LEMMA 2. *We write* $W = \mathrm{ind}_{\mathbb{T}\uparrow\overline{\mathcal{A}}}\chi_0$ *and claim that there is an identification of* $H(W)$ *with* $L^2(\mathcal{A})$ *such that* $R(W)$ *is composed of all operators in* $B(L^2(\mathcal{A}))$ *that are of the matrix form*

$$(a_{y-x}\beta(x,y) \quad (x,y \in \mathcal{A}))$$

for any $a : \mathcal{A} \to \mathbb{C}$, *for which the right-hand side gives rise to a bounded operator on* $L^2(\mathcal{A})$. *Also,* $(R(W))'$ *is the totality of all operators*

$$((a_{y-x}\beta(y,x) \quad (x,y \in \mathcal{A}))$$

where $a : \mathcal{A} \to \mathbb{C}$ *is as before.*

In fact,

a) we have $L^2(\overline{\mathcal{A}}) = L^2(\mathbb{T}) \otimes L^2(\mathcal{A})$.

Let us write $\mathcal{L}$ and $\mathcal{R}$ for the left and right regular representation respectively of $\overline{\mathcal{A}}$ on $L^2(\mathbb{T}) \otimes L^2(\mathcal{A})$. We write $R \subset B(L^2(\mathbb{T}))$ for the von Neumann algebra, generated by the translations of $\mathbb{T}(\subset \overline{\mathcal{A}})$ on itself. We have $R' = R$.

We note that any operator A on $R(\mathcal{R})$ is of the form

$$A = (A_{x,y}; A_{x,y} \in R \quad \text{for } x,y \in \mathcal{A}).$$

b) For fixed $w \in \mathbb{T}$ and $f \in L^2(\mathbb{T})$ we set

$$(K_w f)(u) \equiv f(wu) \quad (u \in \mathbb{T}).$$

For $z \in \mathcal{A}$, we write $L(z) = \mathcal{L}((z,1)^{-1})$. Since $(z,1)(x,u) = (z+x, \beta(z,x)u)$, we have for all $f \in L^2(\mathcal{A})$

$$(L(z)f)(x,u) \equiv (K_{\beta(z,x)})_u f(x+z,u).$$

We have

$$R(\mathcal{R}) = \{A; A = (A_{x,y}), A_{x,y} \in R \quad (x,y \in \mathcal{A})\}$$

and A commutes with $L(z)$ for all $z \in \mathcal{A}$.

c) We note next that $A \in R(\mathcal{R})$ if and only if there is a map $A : \mathcal{A} \to R$ such that

$$A_{x,y} = A_{y-x}K_{\beta(x,y)}$$

assuming that this formula gives rise to a bounded operator on $L^2(\overline{\mathcal{A}})$.

In fact,

1) we have $(AL(z)f)(x,u) = \sum_\delta (A_{x,\delta}(K_{\beta(z,\delta)})_u f(\delta + z, u)$.

2) Similarly we get

$$(L(z)Af)(x,u) = \sum_\delta (K_{\beta(z,x)}A_{x+z,\delta})_u f(\delta,u).$$

We conclude from 1) and 2) that $AL(z) \equiv L(z)A$ if and only if

$$A_{x,\delta-z}K_{\beta(z,\delta)} \equiv K_{\beta(z,x)}A_{x+z,\delta}.$$

Putting $A_y \equiv A_{0,y} \in R$, we get

$$A_{\delta-z}K_{\beta(z,\delta)} \equiv A_{z,\delta}$$

whence, by $z \to x, \delta \to y$ we obtain

$$A_{x,y} \equiv A_{y-x}K_{\beta(x,y)}.$$

Thus $A \in R(\mathcal{R})$ if and only if

$$A = (A_{y-x}K_{\beta(x,y)}; x,y \in \mathcal{A})$$

where $A : \mathcal{A} \to R$ is such that, by the above formula, it gives rise to a bounded operator on $L^2(\overline{\mathcal{A}})$.

d)

1) We note that

$$H(W) = \left\{ F(a,u); a \in \mathcal{A}, u \in \mathbb{T} \right.$$
$$\left. \text{such that } F(a,u) \equiv c_a u \text{ and } \sum_a |c_a|^2 < +\infty \right\}.$$

2) Writing u in place of $(0,u) \in \overline{\mathcal{A}}$, and assuming the Haar measure τ on $\mathbb{T}$ normalized such that

$$\int_{\mathbb{T}} d\tau(u) = 1$$

if $P = \int_{\mathbb{T}} \overline{u}\mathcal{R}(u) \cdot d\tau(u)$, then

$$H(W) = P(L^2(\overline{\mathcal{A}})).$$

3) If we set

$$P_1 = \int_{\mathbb{T}} \overline{u} K_u \cdot d\tau(u)$$

then

$$P = (\delta_{xy} P_1; x, y \in \mathcal{A}).$$

To complete our proof, it is enough to observe that if $A \in R$, then

$$AP_1 = cP_1$$

where c is some constant depending on A. Summing up, we have finished showing the following. Let us put

$$W = \operatorname{ind}_{\mathbb{T}\uparrow\mathcal{A}} \chi_0.$$

Then $R(W)$ is the collection of all operators on $L^2(\mathcal{A})$ of the form

$$(a_{y-x}\beta(x,y); x, y \in \mathcal{A})$$

where the map $a : \mathcal{A} \to \mathbb{C}$ is subject to the sole condition that, by virtue of the above formula, it gives rise to a bounded operator on $L^2(\mathcal{A})$.

LEMMA 3. *With the previous notation, $R(W)$ is a von Neumann algebra of finite type.*

In fact, let us define a linear form f on $R(W)$ by $f(A) = a_0$ if $A = (a_{y-x}\beta(x,y))$. An easy computation shows that

$$f(A \cdot A^*) = \sum_\delta |a_\delta|^2 = f(A^* \cdot A).$$

This being so, it is enough to note that $f(A \cdot A^*) \geqq 0$ and we have here zero if and only if $A = 0$.

We set $\mathcal{A}_0 = \{x; (\beta(x,y))^2 \equiv 1 \text{ for all } y \text{ in } \mathcal{A}\}$. Below we assume that $\mathcal{A}_0$ is finite.

LEMMA 4. *We have $(R(W))^{\mathrm{c}} = R(W|\overline{\mathcal{A}}_0)$.*

In fact, by $\overline{\mathcal{A}}_0 = (\overline{\mathcal{A}})^c$ we have evidently

$$R(W|\overline{\mathcal{A}}_0) \subseteqq (R(W))^c$$

and thus it is enough to show that $R(W)^c \subseteqq R(W|\overline{\mathcal{A}}_0)$. Let us assume that $A \in (R(W))^c$, $A = (a_{y-x}\beta(x,y))$, say. Since $A \in (R(W))'$, by the remark following Lemma 2, there is $B = (b_{y-x}\beta(y,x)) \in (R(W))^c$ such that $A = B$ and thus $(a_{y-x}\beta(x,y)) = (b_{y-x}\beta(y,x))$. This implies $a_z \equiv b_z$ ($z \in \mathcal{A}$). If $a_z \neq 0$, then $\beta(x,z) = \beta(z,x)$ and thus $(\beta(x,z))^2 \equiv 1$ for all $x \in \mathcal{A}$, that is $z \in \mathcal{A}_0$. For $z \in \mathcal{A}$, let us write

$$A(z) = (\delta_{y-x,z}\beta(x,y)).$$

If $z \in \mathcal{A}_0$, then $A(z) \in (R(W))^c$. Since $\mathcal{A}_0$ is finite, we have $A = \sum_{z \in \mathcal{A}_0} a_z A(z)$ and therefore $A \in R(W|\overline{\mathcal{A}}_0)$.

Summing up, we have completed proving that if $\mathcal{A}_0$ is finite, then $R(W|\overline{\mathcal{A}}_0) = (R(W))^c$.

Below we continue to assume that $\mathcal{A}_0$ is finite. Our hypothesis implies that $\overline{\mathcal{A}}_0$ is compact abelian. We recall that $\chi_0((0,u)) \equiv u$ ($u \in \mathbb{T}$). We set

$$E = \{\chi; \chi \in \mathrm{X}(\overline{\mathcal{A}}_0) \text{ such that } \chi|\mathbb{T} \equiv \chi_0\}.$$

For $\chi \in E$, we denote by $P_\chi \in (R(W))^c$ the projection to the eigenspace of χ. By definition, we have $W|\overline{\mathcal{A}}_0 = \bigoplus_{\chi \in E} \chi \cdot P_\chi$.

LEMMA 5. *For $\chi \in E$, let us write W_χ for the part of W in $P_\chi H(W)$. Then W_χ is a factor representation of finite type and we have the central decomposition*

$$W = \bigoplus_{\chi \in E} W_\chi.$$

In fact,

a) by Lemma 4, $(R(W))^c$ is the linear span of $\{P_\chi; \chi \in E\}$.

b) Since, by Lemma 3, $R(W)$ is of finite type, the same is valid for the factor $R(W_\chi)$.

REMARK. The same holds true for $\overline{\chi}_0$ in place of χ_0.

LEMMA 6. *We have* $\dim(P_\chi) = |\mathcal{A}/\mathcal{A}_0|$.

In fact,

a) we define, for $t \in \mathcal{A}$, $e_t \in H(W)$ by $(e_t)_x = \delta_{t,x}$. We recall that for $z \in \mathcal{A}$, by definition

$$A(z) = (\delta_{y-x,z}\beta(x,y)).$$

We note that if $t \in \mathcal{A}$ is given, then

$$A(z)e_t = \beta(t,z)e_{t-z}.$$

Given $x \in \mathcal{A}$, we write M_x for the linear span of $W(\overline{\mathcal{A}}_0)e_x$. Then

$$M_x = \bigoplus_{a \in \mathcal{A}_0} \mathbb{C}e_{x+a}$$

and $H(W) = \bigoplus_x M_x$, where x describes a set of residues modulo $\mathcal{A}_0$.

b) Let $M = M_x$ for some $x \in \mathcal{A}$ fixed. We write $m = \dim(M) = |\mathcal{A}_0|$. We observe that there is a basis in M such that

$$W|\overline{\mathcal{A}}_0 = \operatorname{diag}(\chi_1, \ldots, \chi_m)$$

where the χ's make up E. In fact, clearly: if $\chi_a = \chi_b$ for $a \neq b$, then there is no $\overline{\mathcal{A}}_0$-cyclic vector in M.

c) We conclude from this that $\dim(P_\chi) = |\mathcal{A}/\mathcal{A}_0|$ as claimed in Lemma 6.

LEMMA 7. *With the previous notation, $R(W_\chi)$ is of type $(\mathrm{I}_n, \mathrm{I}_n)$ or $(\mathrm{II}_1, \mathrm{II}_1)$ according to whether $\mathcal{A}/\mathcal{A}_0$ is finite or infinite.*

In fact,

a) let us define the involution S on $L^2(\mathcal{A}) = H(W)$ by $(Sf)(x) \equiv \overline{f(-x)}$. Then, if $A = (a_{y-x}\beta(x,y); x, y \in \mathcal{A})$ then $SAS = (b_{y-x}\beta(y,x); x, y \in \mathcal{A})$, where $b_y = \overline{a}_{-y}$ $(y \in \mathcal{A})$.

b) We have $SP_\chi \equiv P_\chi S$. Hence $A \to SA^*S$ is an involutive anti-isomorphism $R(W_\chi) \to (R(W_\chi))'$.

We recall that our goal is to establish the following statement. Let χ be a character of Γ^c, such that $\chi|\mathcal{T}$ is the identity map of $\mathcal{T}$ onto itself. We put

$$V(\chi) = \mathrm{ind}_{\Gamma^c \uparrow \Gamma} \chi$$

and claim that $V(\chi)$ is a factor representation of Γ of finite type; it is of type one if and only if Γ^c is cofinite in Γ.

We define $\chi_1 \in \mathrm{X}(Z_1)$ by $\chi_1 = \chi|Z_1$ and $\chi_1' \in \mathrm{X}(\mathcal{Z}_1)$ by

$$\chi_1'(x) = \chi_1(x, 1) \quad (x \in \mathcal{Z}_1).$$

Let ψ be a character of $\mathcal{Z}$ such that $\psi|\mathcal{Z}_1 = \chi_1'$. We define $\Psi : \Gamma \to \mathbb{T}$ by: If $a = (x, u)$, then $\Psi(a) = \psi(x)u$. Let us set for $a = (x,u), b = (y,v)$

$$\omega(a,b) = \overline{\alpha(x,y)}.$$

LEMMA 8. *Unexplained notation as above. If a, b are elements of Γ, then $\Psi(a)\Psi(b) = \omega(a,b)\Psi(ab)$.*

In fact, assume that $a = (x,u), b = (y,v)$.

a) We have $\Psi(a)\Psi(b) = \psi(x)u \cdot \psi(y)v = \psi(x+y)uv$.

b) We note that $\omega(a,b)\Psi(ab) = \alpha(x,y)\psi(x+y)\overline{\alpha(x,y)}uv = \psi(x+y)uv = \Psi(a)\Psi(b)$.

We conclude from all this that $\Psi(a)\Psi(b) = \omega(a,b)\Psi(ab)$.

We define the group Γ^e on the set $\Gamma \times \mathbb{T}$ by $(a,u)(b,v) = (ab, \omega(a,b)uv)$. If $a = (x,u), b = ((y,u'),v')$ then we have

$$((x,u),v)((y,u'),v') = (((x+y, \alpha(x,y)u),u'), \overline{\alpha(x,y)}vv').$$

Evidently $\{((z_1,u),v); z_1 \in \mathcal{Z}_1; u, v \in \mathbb{T}\}$ is a central subgroup in Γ^e. We write also Z_1 for the subgroup of Γ^e given by

$$\{((z_1,u),1); z_1 \in \mathcal{Z}, u \in \mathbb{T}\}.$$

Let us define $\Psi_e : \Gamma^e \to \mathbb{T}$ by $\Psi_e((a,u)) = \Psi(a)u$.

LEMMA 9. *Ψ_e is a character of Γ^e.*

In fact,

$$\begin{aligned}\Psi_e((a,u)\cdot(b,v)) &= \Psi_e((ab, \omega(a,b)uv)) \\ &= \Psi(ab)\omega(a,b)\cdot uv = \Psi(a)u\Psi(b) = \Psi_e((a,u))\Psi_e((b,v)).\end{aligned}$$

Hence

$$\Psi_e((a,u)(b,v)) = \Psi_e((a,u))\Psi_e((b,v))$$

and thus Ψ_e is a character of Γ^e.

We define $\Phi : \Gamma^e \to \Gamma^e/Z_1 = M$. Also, we write $T_e = \{(e,u); u \in \mathbb{T}\} \subset \Gamma^e$. If convenient we write u for $(e,u) \in T_e$.

Let R be the right regular representation of M and let R_1 be its part in the subspace with the projection

$$\int_{T_e} uR(u) \cdot d\tau(u) \quad \text{of } L^2(M).$$

The representation $(R_1 \circ \Phi) \otimes \Psi_e$ is identically one on T_e; we shall use it also to denote the corresponding representation of $\Gamma = \Gamma^e/T_e$. This being so, we have

LEMMA 10. *We claim that*

$$(R_1 \circ \Phi) \otimes \Psi_e = \mathrm{ind}_{Z_1 \uparrow \Gamma} \chi_1.$$

In fact, let $\Lambda \subset \mathcal{Z}$ be such that $\mathcal{Z} = \Lambda \mathcal{Z}_1$ setwise. If $z \in \mathcal{Z}$, we write $p(z)$ and $q(z)$ for the components of z in Λ and $\mathcal{Z}_1$, respectively; thus, in particular, $z = p(z) + q(z)$. We have for $\lambda \in \Lambda, z \in \mathcal{Z}$, and $u \in \mathbb{T}$

$$\begin{aligned}(\lambda, 1)(z,u) &= (\lambda + z, \alpha(\lambda, z)u)\\ &= (p(\lambda+z) + q(\lambda+z), \alpha(\lambda,z)u)\\ &= (p(\lambda+z),1)(q(\lambda+z), \alpha(\lambda,z)u).\end{aligned}$$

We conclude from this that if $T = \mathrm{ind}_{Z_1 \uparrow \Gamma} \chi_1$ and $f \in H(T) = L^2(\Lambda)$, then

$$(T(z,u)f)(\lambda) = \chi_1(q(\lambda+z), \alpha(\lambda,z)u)f(p(\lambda+z)).$$

To evaluate the left-hand side, we note that

$$\begin{aligned}((\lambda,1),1)((z,u),v) &= ((\lambda+z, \alpha(\lambda,z)u), \overline{\alpha(\lambda,z)}v\\ &= ((p(\lambda+z)+q(\lambda+z), \alpha(\lambda,z)u), \overline{\alpha(\lambda,z)}v)\\ &= ((p(\lambda+z),1),1)((q(\lambda+z),\alpha(\lambda,z)u)((0,1),\overline{\alpha(\lambda,z)}v)\end{aligned}$$

and thus

$$((R_1 \circ \Phi) \otimes \Psi_e)((z,u),v)f(\lambda) = \psi(z)\alpha(\lambda,z)u \cdot f(p(\lambda+z)).$$

To complete the proof of Lemma 10, we define $V : L^2(\Lambda) \to L^2(\Gamma)$ by $(Vf)(\lambda) = \psi(\lambda)f(\lambda)$. Then we obtain

$$\begin{aligned}(T(z,u)f)(\lambda) &= \chi_1((g(\lambda+z), \alpha(\lambda,z)u))f(p(\lambda+z));\\ (T((z,u)Vf)(\lambda) &= \chi_1((q(\lambda+z),\alpha(\lambda,z)u)\psi(p(\lambda+z))f(p(\lambda+z))\\ &= \psi(\lambda+z)\alpha(\lambda,z)f(p(\lambda+z)).\end{aligned}$$

We have on the other hand

$$V((R_1 \circ \Phi) \otimes \Psi_e)f(\lambda) = \psi(\lambda+z)\alpha(\lambda,z)uf(p(\lambda+z))$$

whence

$$V(R_1 \circ \Phi) \otimes \Psi_e = T \cdot V.$$

In this fashion we obtain that in the sense of unitary equivalence

$$(R_1 \circ \Phi) \otimes \Psi_e = \mathrm{ind}_{Z_1 \uparrow Z} \chi_1.$$

Let us define $\mu : \mathcal{Z} \to \mathcal{Z}/\mathcal{Z}_1 = \mathcal{A}$. There is $\beta : \mathcal{A} \times \mathcal{A} \to \mathbb{T}$, skew and bilinear, such that

$$\beta(\mu(x), \mu(y)) = \overline{\alpha(x, y)}.$$

We define on $\overline{\mathcal{A}} = \mathcal{A} \times \mathbb{T}$ the structure of a group by

$$(a, u)(b, v) = (ab, \beta(a, b)uv).$$

Let us define $\lambda : \Gamma^e \to \overline{\mathcal{A}}$ by

$$\lambda((a, u)) = (\zeta(a), u)$$

where $\zeta((x, u)) = \mu(x)$ $(x \in \mathcal{Z})$.

Lemma 11. *Unexplained notation as above. We claim that λ is a homomorphism from Γ^e onto $\mathcal{A}$.*

In fact, assume that $a, b \in \Gamma$ and $u, v \in \mathbb{T}$. Then we have

$$\begin{aligned}\lambda((a, u))\lambda((b, v)) &= (\zeta(a), u)(\zeta(b), v)\\ &= (\zeta(a)\zeta(b), \beta(\zeta(a), \zeta(b))uv)\\ &= (\zeta(ab), \omega(a, b)uv) = \lambda((a, u)(b, v))\end{aligned}$$

since if $a = (x, u), b = (y, v)$, then

$$\omega(a, b) = \overline{\alpha(x, y)} = \beta(\zeta(a), \zeta(b)).$$

All this being so, we proceed to complete the proof of the main proposition in the following manner. We note first that if $((x, u), v) \in \Gamma^e$, then it lies in the kernel of λ if and only if $x \in \mathcal{Z}_1$ and $v = 1$. We conclude from this that there is an isomorphism $\varepsilon : \overline{\mathcal{A}} \to M$ such that

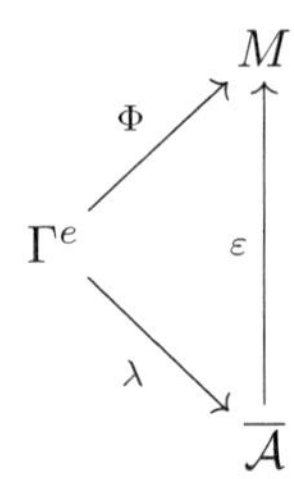

If W_χ $(\chi \in E)$ is as in Lemma 5, then we have $R_1 \circ \varepsilon = \bigoplus_{\lambda \in E} W_\lambda$ and $R_1 \circ \Phi = \bigoplus_{\lambda \in E} (W_\chi \circ \lambda)$.

In this manner we obtain by Lemma 10 that

$$\operatorname{ind}_{Z_1 \uparrow \Gamma} \chi_1 = (R_1 \circ \Phi) \otimes \Psi_e = \bigoplus_{\chi \in E} (W_\chi \circ \lambda) \otimes \Psi_e.$$

The representation corresponding to $(W_\chi \circ \lambda) \otimes \Psi_e$ on $\Gamma = \Gamma^e/T_e$ is a factor representation of the type of W_χ; when restricted to $Z_0 = \Gamma^c$, it coincides with a multiple of a character in $\mathcal{G}$ (cf. Lemma 1). Denoting it by U'_φ and by $\mathcal{G}_1$ the subset of $\mathcal{G}$ formed by the φ's so obtained we get that in the sense of unitary equivalence

$$V = \operatorname{ind}_{Z_1 \uparrow \Gamma} \chi_1 = \bigoplus_{\varphi \in \mathcal{G}_1} U'_\varphi.$$

On the other hand, we have by Lemma 1: $V = \bigoplus_{\varphi \in \mathcal{G}} U_\varphi$ on Γ^c and, on Γ^c, U_φ is identical to a multiple of φ. Hence we conclude that $\mathcal{G} = \mathcal{G}_1$ and $U_\varphi = U'_\varphi$. U_φ

is a factor representation of finite type. Since $\mathcal{A}/\mathcal{A}_0 = \mathcal{Z}/\mathcal{Z}_0 = \Gamma/\Gamma^c$, by Lemma 7 it is of type one if and only if Γ/Γ^c is of a finite order.

REMARK. A similar statement holds true if $\chi|\mathcal{T}$ is the conjugate of the identity map of $\mathcal{T}$ onto itself.

2.2. Proof of Main Proposition

Let Z be a closed subgroup of $\mathbb{R}^n$, and let Γ be a central extension of Z by a 1-torus. We denote by U the centralizer of $(\Gamma_0)^c$ in Γ. Writing $\mathcal{T}$ for the canonical torus in Γ, we form the set

$$\mathcal{F} = \{\chi; \chi \in \mathrm{X}(U^c) \text{ such that } \chi|\mathcal{T} \text{ is the identity map } \mathcal{T} \to \mathcal{T}\}.$$

Given $\chi \in \mathcal{T}$, we set

$$V(\chi) = \mathrm{ind}_{U^c \uparrow \Gamma}\, \chi.$$

With this notation, we have the following.

PROPOSITION.
1) *$V(\chi)$ is a semifinite factor representation.*
2) *$V(\chi)$ is of type one if and only if*

$$U^c\Gamma_0 \textit{ is of finite index in } U.$$

3) *If χ and χ' are in $\mathcal{F}$, $V(\chi)$ is quasi-equivalent to $V(\chi')$ if and only if $\Gamma\chi = \Gamma\chi'$.*

We start by recalling (cf. [**AM**], p. 188) that there is a skew symmetric bilinear form $B : Z \times Z \to \mathbb{R}$ such that if we set $\beta(x, y) = \exp((i/2)B(x, y))$ $(x, y \in Z)$, then Γ is the collection of all pairs (x, u) $(x \in Z, u \in \mathbb{T})$ with the law of composition $(x, y \in Z; u, v \in \mathbb{T})$

$$(x, u)(y, v) = (x + y, \beta(x, y)uv).$$

Given $\chi \in \mathrm{X}(U^c)$ and $a \in \Gamma$ fixed, we define

$$(a\chi)(g) = \chi(a^{-1}ga) \quad (g \in U^c).$$

This being so, we have the

LEMMA 1. *Given any χ in $\mathcal{F}$, its stabilizer in Γ is equal to U.*

In fact, if $a \in U$, then we have trivially $a\chi = \chi$. Suppose now that $a\chi = \chi$. Since $[\Gamma, \Gamma] \subseteqq \mathcal{T}$, and $\chi|\mathcal{T}$ is the identity map, we have for all $g \in (\Gamma_0)^c : [a, g] = e$, implying $a \in U$.

Before proceeding, let us note that $V(\chi)|U^c$ is quasi-equivalent to $\bigoplus_{\gamma \in \Gamma} \gamma\chi$. Therefore, if $V(\chi)$ is quasi-equivalent to $V(\chi')$, then $\Gamma\chi = \Gamma\chi'$. Conversely, the last relation implies that $V(\chi)$ and $V(\chi')$ are unitarily equivalent. Let us put $U(\chi) = \mathrm{ind}_{U^c \uparrow U}\, \chi$. We have then by induction through stages $V(\chi) = \mathrm{ind}_{U \uparrow \Gamma}\, U(\chi)$. To complete our proof it is enough to show that $U(\chi)$ is a semifinite factor representation, which is of type one if and only if $U^c \cdot \Gamma_0$ is of finite index in U. In fact, if so, since $U(\chi)|U^c$ is a multiple of χ, by Lemma 1, $V(\chi)$ is a factor representation of the type of $U(\chi)$ (cf. [**AM**], Theorem 6, p. 58).

Given a subset W of Z, whenever convenient, we shall write W also for the subset $\{(w,1); w \in W\}$ of Γ. We put

$$W_B^{\perp} = \{z; z \in Z, B(z,w) = 0 \text{ for all } w \in W\}.$$

LEMMA 2. *If Γ_1 is the centralizer of Γ_0 in Γ, we have*

$$\Gamma_1 = (Z_0)_B^{\perp} \cdot \mathcal{T}.$$

In fact, an easy computation shows that for any $(x,u),(y,v)$ in Γ, we have

$$(x,u)^{-1} \cdot (y,v)(x,u) = (y,v)(0,(\beta(y,x))^2).$$

Therefore (y,v) lies in Γ_1 if and only if $(\beta(y,z))^2 \equiv 1$ for all $z \in Z_0$, that is, y lies in $(Z_0)_B^{\perp}$.

Let Z_1 be the radical of $B|(Z_0 \times Z_0)$, that is, $Z_1 = (Z_0)_B^{\perp} \cap Z_0$.

LEMMA 3. *We have $U = (Z_1)_B^{\perp} \cdot \mathcal{T}$.*

In fact, we have $(\Gamma_0)^{\mathrm{c}} = Z_1 \cdot \mathcal{T}$. Therefore, (a,u) is in U if and only if, for all $z \in Z_1$, $(\beta(a,z))^2 \equiv 1$, which is equivalent to $(a,u) \in (Z_1)_B^{\perp} \cdot \mathcal{T}$.

Below we make repeated use of the relation

$$(Z_0)_B^{\perp} + Z_0 = (Z_1)_B^{\perp}.$$

LEMMA 4. *With the previous notation we have $U = \Gamma_0 \cdot \Gamma_1$.*

This follows at once from Lemmas 2 and 3 along with the above observation.

Since Z_1 is the radical of $B|(Z_0 \times Z_0)$ it is the connected component of zero in $(Z_0)_B^{\perp}$. Hence there is a discrete subgroup Σ of Z such that

$$(Z_0)_B^{\perp} = Z_1 \oplus \Sigma$$

and

$$(Z_1)_B^{\perp} = Z_0 + (Z_0)_B^{\perp} = Z_0 + \Sigma.$$

We have evidently setwise

$$U = \Gamma_0 \cdot \Sigma.$$

Let Ψ be the projection $U \to \Sigma$. We define $\omega : U \times U \to \mathbb{T}$ by $\omega(a,b) = \overline{\beta(\Psi(a),\Psi(b))}$. Let U^e be the set of all pairs (a,u) $(a \in U, u \in \mathbb{T})$ with the law of composition

$$(a,u)(b,v) = (ab, \omega(a,b)uv).$$

LEMMA 5. *With the above notation, we claim that U^e is a group.*

In fact, we recall (cf. [**AM**], p. 21) that the multiplier condition is $(a,b,c \in U)$

$$\omega(bc,a)\omega(b,c) \equiv \omega(b,ca)\omega(c,a).$$

Let us put

$$r = \Psi(a), \quad s = \Psi(b), \quad t = \Psi(c) \quad (r,s,t \in \Sigma);$$

we have: $\omega(bc,a) = \overline{\beta(\Psi(b,c),\Psi(a))}$. But also $\Psi(b,c) = s+t$, $\Psi(a) = r$. Therefore

$$\omega(bc,a) = \exp((i/2)B(s+t,r)) \text{ etc.}$$

Hence all we have to know is

$$B(s+t,r) + B(s,t) = B(s,r+t) + B(t,r),$$

which is a valid statement.

If V is a subgroup of U, we shall write V^e for its complete inverse image in U^e. Given $S \subset U$, we shall write sometimes again S for $\{(s,1); s \in S\} \subset U^e$. For instance, $\sigma \in \Sigma$ can stand also for $((\sigma,1),1)$. We denote by T_e the subset $\{(e,u); |u| = 1\}$ of U^e.

We leave to the reader to verify that if $(b,u) \in U^e$ and $(a,1) \in \Gamma_0$, then

$$(b,u)(a,1)(b,u)^{-1} = (bab^{-1},1).$$

We have also setwise: $\Gamma_1^e = \Sigma \cdot (\Gamma_0)^c T_e$. We write χ_0 for $\chi|(\Gamma_0)^c$ ($\chi \in \mathcal{F}$) and define $\chi^e : \Gamma_1^e \to \mathbb{T}$ by

$$\chi^e(\sigma a u) = \chi_0(a)u.$$

LEMMA 6. *With the above notation, χ^e is a character of Γ_1^e.*

In fact, assume that $\sigma, \tau \in \Sigma$, $a, b \in (\Gamma_0)^c$ and $u, v \in \mathbb{T}$. We have: $\sigma a u \cdot \tau b v = \sigma\tau \cdot ab \cdot uv$. But

$$\sigma\tau = ((\sigma,1),1)((\tau,1),1) = ((\sigma+\tau,\beta(\sigma,\tau)),\overline{\beta(\sigma,\tau)}).$$

We write $\alpha(\sigma,\tau) = ((0,\beta(\sigma,\tau)),1)$. We have then

$$\sigma a u \cdot \tau b v = (\sigma+\tau)\alpha(\sigma,\tau) \cdot ab \cdot uv \cdot \overline{\beta(\sigma,\tau)}$$

from where we conclude that

$$\begin{aligned}\chi^e(\sigma a u \cdot \tau b v) &= \chi_0(\alpha(\sigma,\tau) \cdot ab)uv \cdot \overline{\beta(\sigma,\tau)} \\ &= \chi_0(ab)uv = \chi_0(a)u \cdot \chi_0(b)v \\ &= \chi^e(\sigma a u) \cdot \chi^e(\tau b v)\end{aligned}$$

which implies $\chi^e \in \mathrm{X}(\Gamma_1^e)$.

Let γ be the Lie algebra of Γ. We can assume $e_0 \in \gamma$ to have been chosen such that if $z \in Z_0$ and $c \in \mathbb{R}$, then $\exp(z + ce_0) = (z, \exp(ic))$. One verifies easily that if $t, t' \in Z_0$ and $a, b \in \mathbb{R}$, then we have

$$[t + ae_0, t' + be_0] = B(t,t')e_0.$$

Also, $[(x,u),(y,v)] = \exp(B(x,y)e_0)$.

Let $\mathfrak{h}$ be a maximal abelian subalgebra in γ. Evidently, $\mathfrak{h}$ is an ideal in γ. We write H for the corresponding connected subgroup of Γ. If d is an element of the dual γ^* of γ, and (e_0, d) is an integral multiple of 2π, then there is a character χ_d of H such that $\chi_d(\exp(l)) = \exp[i(l,d)]$ ($l \in \mathfrak{h}$). We recall that, by what we saw above, $\Gamma_1^e = \Sigma \cdot (\Gamma_0)^c \cdot \Gamma_e$ setwise. Let χ be an element of $\mathcal{F}$. By Lemma 6, if we define for $\sigma \in \Sigma_1$, $a \in (\Gamma_0)^c$ and $u \in T_e : \chi^e(\sigma a u) = \chi_0(a)u$, then χ^e is a character of Γ_1^e.

LEMMA 7. *With the above notation, let d be an element of the dual of γ such that $\chi_d|(\Gamma_0)^c = \chi|(\Gamma_0)^c$. If we form*

$$V_d = \mathrm{ind}_{H\uparrow\Gamma_0}\, \chi_d$$

then V_d is an irreducible representation of Γ_0. Let us set

$$W = \mathrm{ind}_{\Gamma_1^e \uparrow U^e}\, \chi^e.$$

Then there is a unitary representation ρ of U^e, to which W is quasi-equivalent, and the restriction of which to Γ_0 is unitarily equivalent to V_d.

In fact,

a) we form $H' = \{(h, 1) \in U^e; h \in H\}$. Then H' commutes with every element of Γ_1^e, and $\Gamma_1^e \cdot H'$ is a closed subgroup of U^e. We denote it by K.

b) There is an element χ'_d of $\mathrm{X}(K)$ such that $\chi'_d|\Gamma_1^e = \chi^e$ and $\chi'_d|H' = \chi_d$. To see this, it is enough to note that $\Gamma_1^e \cap H' = (\Gamma_0)^c$ and $\chi_d|(\Gamma_0)^c = \chi|(\Gamma_0)^c$.

c) Let us write λ for the canonical morphism of K onto $K/\Gamma_1^e, = L$ (say). We have

$$L = K/\Gamma_1^e = \Gamma_1^e H'/\Gamma_1^e = H'/(H' \cap \Gamma_1^e) = H'/(\Gamma_0)^c.$$

We conclude from this that L is a vector group. If $d\chi$ is the element of the Haar measure on $\mathrm{X}(L)$, then we have (cf. [**Mack Jer**], Lemma 1, p. 325)

$$S = \int_{\mathrm{X}(L)}^{\oplus} \chi \cdot d\chi.$$

This being so, we have

$$\mathrm{ind}_{\Gamma_1^e \uparrow K}\, \chi^e = (S \circ \lambda) \otimes \chi'_d.$$

d) Let ω be an element of $\mathrm{X}(L)$ and write $\varphi = \omega \circ \lambda$. We claim that there is an element m of Γ_0, such that $\varphi\chi'_d = m\chi'_d$. In fact, since $K = \Gamma_1^e \cdot H' = \Sigma \cdot H' \cdot T_e$, and since Σ commutes with Γ_0, it is enough to show that there is an element m of Γ_0 such that $m\chi_d = (\varphi|H)\chi_d$. There is an element f of γ^* such that $\chi_d(\varphi|H) = \chi_f$. Therefore, it suffices to prove that $\Gamma_0 d = \Gamma_0 f$. But, by $\varphi|(\Gamma_0)^c \equiv 1$ we have $\chi_d|(\Gamma_0)^c = \chi_f|(\Gamma_0)^c$. Hence the desired conclusion follows from $[\gamma, \gamma] \subseteqq \mathbb{R}e_0$.

e) Let us form $\rho = \mathrm{ind}_{K \uparrow U^e}\, \chi'_d$. We claim that W is quasi-equivalent to ρ. In fact, we have by the chain rule

$$W = \mathrm{ind}_{\Gamma_1^e \uparrow U^e}\, \chi^e = \mathrm{ind}_{K \uparrow U^e}\Big(\mathrm{ind}_{\Gamma_1^e \uparrow K}\, \chi^e\Big).$$

We have by the end of c)

$$\mathrm{ind}_{\Gamma_1^e \uparrow K}\, \chi^e = (S \circ \lambda)\chi'_d = \int_{\mathrm{X}(L)} (\omega \circ \lambda)\chi'_d\, d\omega$$

which implies

$$W = \int_{\mathrm{X}(L)}^{\oplus} \mathrm{ind}_{K \uparrow U^e}(\omega \circ \lambda)\chi'_d \cdot d\omega.$$

But, by d), $\mathrm{ind}_{K \uparrow U^e}(\omega \circ \lambda)\chi'_d$ is unitarily equivalent to $\mathrm{ind}_{K \uparrow U^e}\, \chi'_d = \rho$, from where we conclude that

$$W = \int_{\mathrm{X}(L)}^{\oplus} \rho \cdot d\omega$$

which is quasi-equivalent to ρ.

f) By what we have seen it is enough to show that $\rho|\Gamma_0$ is unitarily equivalent to V_d. To establish this point, we shall use the following special case of [**Mackann I**], Theorem 12.1, p. 127. Assume that G_1 and G_2 are closed subgroups of the separable locally compact group G such that $G = G_1 \cdot G_2$. For a given $\chi \in \mathrm{X}(G_1)$ let us write $\chi' = \chi|(G_1 \cap G_2)$. Then we have

$$\Big(\mathrm{ind}_{G_1 \uparrow G}\, \chi\Big)|G_2 = \mathrm{ind}_{G_1 \cap G_2 \uparrow G_2}\, \chi'.$$

Let us take $G = U^e, G_1 = K, G_2 = \Gamma_0$, and $\chi = \chi'_d$. Then we have

$$K\Gamma_0 = U^e, \qquad K \cap \Gamma_0 = H', \qquad \chi'_d | H' = \chi_d$$

which implies that

$$\rho|\Gamma_0 = \mathrm{ind}_{K\uparrow U^e}(\chi'_d|\Gamma_0) = \mathrm{ind}_{H'\uparrow\Gamma_0}\,\chi_d = V_d$$

and thus $\rho|\Gamma_0$ is unitarily equivalent to V_d.

Let us consider the set $\{(\sigma, u); \sigma \in \Sigma, u \in \mathbb{T}\}$ with the law of composition

$$(\sigma, u)(\tau, v) = (\sigma + \tau, \overline{\beta(\sigma, \tau)}uv).$$

We denote by M the group so arising.

We recall that we have the direct sum decomposition $(Z_1)^{\perp}_B = Z_0 + \Sigma$. Let ψ be the projection onto the second summand, and define the map $\Phi : U^e \to M$ by $\Phi((x, u), v) = (\psi(x), v)$. This being so we have

LEMMA 8. *We claim that Φ is a surjective morphism $U^e \to M$.*

In fact, let $x, y \in (Z_1)^{\perp}_B$ and $u, u', v, v' \in \mathbb{T}$ be given. We have in U^e

$$((x, u), u')((y, v), v') = ((x + y, \dots), \omega((x, u), (y, v))u'v').$$

But $\omega((x, u), (y, v)) = \overline{\beta(\psi(x), \psi(y))}$ and therefore $((x, u), u')((y, v), v') = ((x + y, \dots), \overline{\beta(\psi(x), \psi(y))}u'v')$. We obtain in this manner

$$\begin{aligned}\Phi(((x, u), u') \cdot ((y, v), v')) &= (\psi(x + y), \overline{\beta(\psi(x), \psi(y))}u'v') \\ &= (\psi(x), u')(\psi(y), v') = \Phi((x, u), u')\Phi((y, v), v')\end{aligned}$$

which proves our lemma. Since, by virtue of our notational convention, Γ_0 stands also for $\{((x, u), 1); z \in Z_0, u \in \mathbb{T}\}$, we have $\Phi : U^e \to U^e/\Gamma_0 = M$.

LEMMA 9. *The complete inverse image of M^c through Φ is equal to $(U^c)^e \cdot \Gamma_0$.*

In fact,

a) Since we have in M

$$(\sigma, u)(\tau, v) = (\sigma + \tau, \overline{\beta(\sigma, \tau)}u \cdot v),$$

(σ, u) belongs to M^c if and only if $(\beta(\sigma, \tau))^2 \equiv 1$ for all $\tau \in \Sigma$. We use the symbol Σ_1 for the set of all such σ's.

b) An easy verification shows that $U^e = \Sigma \cdot \Gamma_0 \cdot \mathbb{T}_e$ setwise.

c) Clearly, given $\sigma \in \Sigma$ and $a \in \Gamma_0$, the product σa is in U^c if and only if $\sigma \in \Sigma_1, a \in (\Gamma_0)^c$. We conclude from this that $(U^c)^e = \Sigma_1 \cdot (\Gamma_0)^c \cdot T_e$ and thus

$$(U^c)^e \cdot \Gamma_0 = \Sigma_1 \cdot \Gamma_0 \cdot T_e = \Phi^{-1}(M^c)$$

as claimed in our lemma.

Let η be a character of M^c such that $\eta(0, u) \equiv \overline{u}$ $(u \in \mathbb{T})$. We form $A = \mathrm{ind}_{M^c\uparrow M}\,\eta$ and recall that by the Proposition of §2.1, A is a factor representation of finite type; it is of type one if and only if M/M^c is finite. We have by Lemmas 8 and 9

$$\Phi^{-1}(M) = U^e, \qquad \Phi^{-1}(M^c) = (U^c)^e \cdot \Gamma_0$$

and thus M/M^c is equal to $U/(U^c \cdot \Gamma_0)$. We write $B = A \circ \Phi$ and note that B is a factor representation of finite type of U^e. It is of type one if and only if $U^c \cdot \Gamma_0$ is of finite index in U.

We recall that we defined W as $\operatorname{ind}_{\Gamma_1^e \uparrow U^e} \chi^e$ and showed in Lemma 7 that W is quasi-equivalent to a representation ρ, the restriction of which to Γ_0 is irreducible. Evidently, the restriction of W to T_e is equal to the identity map of the circle group onto itself. Thus if we form $B \otimes W$, this is identically one on $T_e \subset U^e$. Hence it gives rise to a representation of $U = U^e/T_e$, which we shall often denote again by $B \otimes W$. Immediate implications of this construction are:

1) If $\mathfrak{B}$ is the full ring of $H(\rho)$, then $R(B \otimes W)$ is a multiple of $R(B) \otimes \mathfrak{B}$.

2) $R(B \otimes W)$ is a semifinite factor; it is of type one if and only if $U/U^c \cdot \Gamma_0$ is finite.

From here we proceed to the demonstration of the Proposition of this section as follows. We recall that $U(\chi) = \operatorname{ind}_{U^c \uparrow U} \chi$. It is enough to prove that there is a choice of η such that $B \otimes W$ is unitarily equivalent to $U(\chi)$. To this end we shall employ the following special case of [**Mackact**], 12.2, p. 128. Let G_1 and G_2 be closed subgroups of the separable locally compact group G such that $G = G_1 \cdot G_2$. Assuming that $\chi_d \in \mathrm{X}(G_j)$ $(j = 1, 2)$ are given characters, we put $\chi' = \chi_1 \cdot \chi_2 | (G_1 \cap G_2)$. This being so we have in the sense of unitary equivalence

$$\left(\operatorname{ind}_{G_1 \uparrow G} \chi_1\right) \otimes \left(\operatorname{ind}_{G_2 \uparrow G} \chi_2\right) = \operatorname{ind}_{G_1 \cap G_2 \uparrow G} \chi'.$$

Below $\chi \in \mathrm{X}(U^c)$ will be as in our principal statement. We define the map $\eta : M^c \to \mathbb{T}$ by $\eta(\sigma, u) \equiv \chi(\sigma, 1)\overline{u}$ $(\sigma \in \Sigma_1, u \in \mathbb{T})$. We claim that η is a character of M^c. We have, in fact, for all $\sigma, \tau \in \Sigma_1$ and $u, v \in \mathbb{T}$: $\eta((\sigma, u))\eta((\tau, v)) = \chi(\sigma, 1)\chi(\tau, 1)\overline{u} \cdot \overline{v}$. But

$$\begin{aligned}\chi(\sigma, 1)\chi(\tau, 1) &= \chi((\sigma, 1)(\tau, 1)) \\ &= \chi(\sigma + \tau, \beta(\sigma, \tau)) = \chi(\sigma + \tau, 1)\beta(\sigma, \tau).\end{aligned}$$

We have thus $\eta(\sigma, u)\eta(\tau, v) = \chi(\sigma + \tau, 1)\beta(\sigma, \tau)\overline{u}\,\overline{v}$.

On the other hand, we get in M^c

$$(\sigma, u)(\tau, v) = (\sigma + \tau, \overline{\beta(\sigma, \tau) u \cdot v})$$

and hence

$$\eta((\sigma, u)(\tau, v)) = \chi(\sigma + \tau, 1)\beta(\sigma, \tau)\overline{u}\,\overline{v}.$$

In this fashion η is in $\mathrm{X}(M^c)$ and clearly $\eta(0, u) \equiv \overline{u}$.

All this being said, we propose to specify the ingredients in the above reminder as follows:

1) $G_1 = \Phi^{-1}(M^c)$ $(= \Sigma_1 \cdot \Gamma_0 \cdot T_e)$ and $\chi_1 = \eta \circ \Phi$;

2) $G_2 = \Gamma_1^0$ $(= \Sigma\Gamma_0^c T_e)$ and $\chi_2 = \chi^e$.

We have in this manner

$$G_1 \cdot G_2 = \Sigma_1 \cdot \Gamma_0 \cdot T_e \cdot \Sigma(\Gamma_0)^c \cdot T_e = \Sigma \cdot \Gamma_0 \cdot T_e = U^e$$

and

$$G_1 \cap G_2 = \Sigma_1 \cdot (\Gamma_0)^c \cdot T_e.$$

We have, if $\sigma \in \Sigma_1, a \in (\Gamma_0)^c$ and $u \in T_e$: $\chi_1(\sigma a u) = \chi(\sigma, 1)\overline{u}$ and $\chi_2(\sigma a u) = \chi(a)u$. Hence on $U^c : \chi_1\chi_2 = \chi$. We have also

$$B = A \circ \Phi = \left(\operatorname{ind}_{M^c \uparrow M} \eta\right) \circ \Phi = \operatorname{ind}_{\Phi^{-1}(M^c) \uparrow U^e}(\eta \circ \Phi) = \operatorname{ind}_{G_1 \uparrow G} \chi,$$

and hence $B = \operatorname{ind}_{G_1 \uparrow G} \chi_1$ and

$$W = \operatorname{ind}_{\Gamma_1^e \uparrow U^e} \chi^e = \operatorname{ind}_{G_2 \uparrow G} \chi_2.$$

From here we conclude that

$$\begin{aligned} B \otimes W &= \Big(\operatorname{ind}_{G_1 \uparrow G} \chi_1\Big) \otimes \Big(\operatorname{ind}_{G_2 \uparrow G} \chi_2\Big) \\ &= \operatorname{ind}_{G_1 \cap G_2 \uparrow G} \chi^e = \operatorname{ind}_{U^e \uparrow U} \chi \end{aligned}$$

and thus finally on $U = U^e/T_e : B \otimes W = U(\chi)$ which completes the proof of our Proposition.

2.3. Primitive ideals of class one

We continue to assume that Γ is a central extension by a 1-torus of a closed subgroup Z of $\mathbb{R}^n$ and shall employ the notation of §2.2. We start by recalling (cf. [**Dal**], Corollaire 3, p. 100) that the kernel of a factor representation of a separable C^*-algebra is always a primitive ideal. We denote by $\operatorname{Prim}(\Gamma)$ the primitive ideal space of the group C^*-algebra of Γ. By §2.2, given $\chi \in \mathcal{F}$, $\ker(V(\chi))$ belongs to $\operatorname{Prim}(\Gamma)$; we shall denote it by $J(\chi)$. Given a factor representation U of Γ, $U|\mathcal{T}$ depends only on $\ker(U)$. Hence it makes sense to speak of the primitive ideals of class one of $\operatorname{Prim}(\Gamma)$. We denote by $\operatorname{Prim}_1(\Gamma)$ their totality. In particular, $J(\chi)$ belongs to $\operatorname{Prim}_1(\Gamma)$. The rest of this chapter will be devoted to proving that $\chi \mapsto J(\chi)$ is a surjective map from $\mathcal{F}$ onto $\operatorname{Prim}_1(\Gamma)$ and that we have $J(\chi_1) = J(\chi_2)$ if and only if the closures of $\Gamma\chi_1$ and $\Gamma\chi_2$ in $\mathcal{F}$ coincide. We shall start by establishing the second assertion.

Below we shall write $\mathfrak{G}$ for the set of all characters of U^c that are identically one on $\mathcal{T}$.

LEMMA 1. *There is a closed subgroup $\mathfrak{H}$ of $\mathfrak{G}$ such that we have for any χ in $\mathcal{F} : \overline{\Gamma\chi} = \mathfrak{H}\chi$.*

In fact, let χ be a fixed element of $\mathcal{F}$. Since $[\Gamma, \Gamma] \subseteqq \mathcal{T}$, and $\chi|\mathcal{T}$ is the identity map, given χ in Γ, there is an element φ_γ of $\mathfrak{G}$, independent of χ such that $\gamma\chi = \varphi_\gamma \cdot \chi$. Let us write $\mathfrak{H}$ for the closure of $\{\varphi_\gamma; \gamma \in \Gamma\}$ in $\mathfrak{G}$. This being so we claim that, given $\chi \in \mathcal{F}$, the closure of $\Gamma\chi$ in $\mathcal{F}$ is equal to $\mathfrak{H}\chi$. In fact, clearly $\Gamma\chi \subseteqq \mathfrak{H}\chi$ and thus $\overline{\Gamma\chi} \subseteq \mathfrak{H}\chi$. But, by construction, $\Gamma\chi$ is dense in $\mathfrak{H}\chi$ and, in this manner, $\mathfrak{H}\chi = \overline{\Gamma\chi}$.

LEMMA 2. *Suppose that $F \subset E$ are closed subgroups in $\mathbb{R}^n$, and that $(\chi_n) \subset \widehat{F}$ is such that $\chi_n \to 1$ in $\widehat{F}$. Then there is a sequence (ψ_n) in $\widehat{E}$ such that $\psi_n|F \equiv \chi_n$ and $\psi_n \to 1$.*

In fact, we recall (cf. [**Hochs**], p. 34) that if V is a vector subgroup and Q a discrete subgroup of $\mathbb{R}^n$ such that $F = V + Q$ (direct sum) then there is a vector subgroup U and a discrete subgroup P such that E is equal to the direct sum $U+P$, and $V \subseteqq U, Q = (Q \cap U) + (Q \cap P)$. In this manner the proof is reduced to one of the following cases, each of which is trivial:

1) $F = V$;
2) $F = Q$ and $Q \subseteqq U$;
3) $F = Q$ and $Q \subseteqq P$.

LEMMA 3. *We claim that the map $\chi \mapsto J(\chi)$ from $\mathcal{F}$ into* $\operatorname{Prim}(\Gamma)$ *is continuous.*

In fact, let us assume that the sequence $(\chi_n) \subset \mathcal{F}$ tends to χ_0. We can write $\chi_n \equiv \varphi_n \chi_0$, where (φ_n) tends to one. By Lemma 2, there is a sequence (Φ_n) of characters of Γ such that $\Phi_n | U^{\mathrm{c}} \equiv \varphi_n$ and $\Phi_n \to 1$. In this fashion we have, in particular, $V(\chi_n) \equiv \Phi_n(V(\chi_0))$. Clearly, $V(\chi_0)$ is weakly contained in any subsequence of $(V(\chi_n))$ and thus $V(\chi_0)$ is in the closure of any subsequence of $(J(\chi_n))$. Therefore, $J(\chi_n)$ tends to $J(\chi_0)$ and hence $\chi \mapsto J(\chi)$ is continuous from $\mathcal{F}$ into $\mathrm{Prim}(\Gamma)$.

LEMMA 4. *Assume that χ_1 and χ_2 in $\mathcal{F}$ are such that $J(\chi_1) \supseteqq J(\chi_2)$. Then the closures of $\Gamma\chi_1$ and $\Gamma\chi_2$ coincide.*

In fact, the assumption implies that $V(\chi_1)$ is weakly contained in $V(\chi_2)$. Hence $V(\chi_1)|U^{\mathrm{c}}$ is weakly contained in $V(\chi_2)|U^{\mathrm{c}}$. But $V(\chi_j)|U^{\mathrm{c}}$ is weakly equivalent to $\overline{\Gamma\chi_j}$ $(j=1,2)$. Therefore $\overline{\Gamma\chi_1} \subseteqq \overline{\Gamma\chi_2}$. But by Lemma 1, $\overline{\Gamma\chi_j} = \mathfrak{H}\chi_j$ $(j=1,2)$ and hence $\overline{\Gamma\chi_1} = \overline{\Gamma\chi_2}$.

The following lemma is the principal assertion of this section.

LEMMA 5. *Assume that $\chi_j \in \mathcal{F}$ $(j=1,2)$. Then the closures of $\Gamma\chi_j$ $(j=1,2)$ coincide if and only if $J(\chi_1) = J(\chi_2)$.*

In fact, if $J(\chi_1) = J(\chi_2)$, then $J(\chi_1) \supseteqq J(\chi_2)$. Then, by Lemma 4, the closures of $\Gamma\chi_j$ $(j=1,2)$ coincide. Hence it is enough to show that, assuming the second relation, $J(\chi_1) = J(\chi_2)$. The hypothesis implies $\ker(V(\chi_1)|U^{\mathrm{c}}) = \ker(V(\chi_2)|U^{\mathrm{c}}) \Rightarrow \ker \mathrm{ind}_{U^{\mathrm{c}}\uparrow\Gamma}(V(\chi_1)|U^{\mathrm{c}}) = \ker \mathrm{ind}_{U^{\mathrm{c}}\uparrow\Gamma}(V(\chi_2)|U^{\mathrm{c}})$, and on the other hand, $\mathrm{ind}_{U^{\mathrm{c}}\uparrow\Gamma}(V(\chi)|U^{\mathrm{c}}) \approx V(\chi)$.

For later use we note

LEMMA 6. *Assume that for J in $\mathrm{Prim}_1(\Gamma)$ there are elements χ_1, χ_2 in $\mathcal{F}$ such that $J(\chi_1) \supseteqq J \supseteqq J(\chi_2)$. Then there is a χ in $\mathcal{F}$ such that $J = J(\chi)$.*

In fact, the assumption implies that $J(\chi_1) \supseteqq J(\chi_2)$. Then, by Lemma 4, the closures of $\Gamma\chi_j$ $(j=1,2)$ coincide. Hence, by Lemma 5, $J(\chi_1) = J = J(\chi_2)$.

2.4. Surjectivity; first step

The objective of this section is the following

PROPOSITION. *Assume that J is some element of $\mathrm{Prim}_1(\Gamma)$. Then there is a χ in $\mathcal{F}$ such that $J(\chi) \subseteqq J$.*

In fact,

a) Below we shall write: $G = \Gamma, K = U^{\mathrm{c}}$, and $\pi : G \to G/K = A$. Assume that T is a factor representation of G such that $\ker(T) = J$ (as given). Since T is of class one, there is a Borel measure μ on $\mathcal{F}$ such that

$$T|K = \int_{\mathcal{F}}^{\oplus} \rho(\chi) \cdot d\mu(\chi)$$

where $\rho(\chi)$ is quasi-equivalent to χ.

Before proceeding we note that since T is a factor representation, μ is ergodic with respect to the action of G on $\mathcal{F}$. Given χ in $\mathcal{F}$, we write again: $V(\chi) = \mathrm{ind}_{K\uparrow G}\,\chi$. By 2.2, $V(\chi)$ is a factor representation. There is a positive integer n such that $\rho(\chi) \equiv n\chi$ ($\chi \in \mathcal{F}$). We define $T' = \mathrm{ind}_{K\uparrow G}(T|K)$. Putting $V'(\chi) \equiv nV(\chi)$ we have

$$T' = \int_{\mathcal{F}}^{\oplus} V'(\chi) \cdot d\mu(\chi).$$

We claim that T' is a factor representation of G. This follows from the following.

LEMMA. *Assume that G is separable and locally compact. Let (X, μ) be an ergodic Borel G-space. Suppose that $V(x)$ is a μ-measurable field of factor representations such that, if $Gx = Gx'$, $V(x)$ is unitarily equivalent to $V(x')$. We put*

$$V = \int_{X}^{\oplus} V(\chi) \cdot d\mu(\chi).$$

This being so, we claim that V is a factor representation of G.

1) Assume that A is a bounded, decomposable operator, commuting with V. We can assume first that $A(x)$ commutes with $V(x)$ almost everywhere with respect to μ, and next, by modification on a set of μ-measure zero, that $A(x)$ commutes with $V(x)$ everywhere in X.

2) We assume next that $A \in R(V)$ (in which fashion $A \in (R(V))^{\mathrm{c}}$). Then there is a sequence (A_n) of linear combinations of $(V(a); a \in G)$ such that A_n tends to A strongly. Passing, if necessary, to a subsequence, we can assume that $A_n(x)$ tends to $A(x)$ strongly, almost everywhere with respect to μ. The set of points in x, giving rise to a converging sequence $(A_n(x))$ is G-invariant, and its complement is of μ-measure zero. We conclude by all this that $A(x)$ is in $R(V(x))$ almost everywhere and hence, since $R(V(x))$ is a factor, $A(x) \equiv \varphi(x) \cdot I$, the exception being as above.

3) Since, if $Gx = Gx'$, $V(x)$ is unitarily equivalent to $V(x')$, we have $\varphi(x) = \varphi(x')$ for $Gx = Gx'$, the exceptions being as in 2). Hence, by ergodicity, there is a constant c such that $\varphi(x) \equiv c$, almost everywhere with respect to μ.

Summing up, if A is in $(R(V))^{\mathrm{c}}$, then it is a constant multiple of the identity operator, and hence $R(V)$ is a factor.

b) Since $T' = \int_{\mathcal{F}}^{\oplus} V'(\chi) \cdot d\mu(\chi)$, and μ is ergodic, our lemma implies that T' is a factor representation. We have also (cf. [**Dal**], p. 100) that $\ker(T') \equiv \ker(V(\chi)) \equiv J(\chi)$ almost everywhere. Therefore, putting $I = \ker(T')$, it will be enough to show that $J \supseteq I$.

Below we shall be inspired by [**EH**].

c) We shall write $M(G)$ for the set of all Radon measures on G, and $D(G)$ for the collection of all positive definite measures; similarly for A in place of G. For $\mu, \nu \in D(G)$ we shall write $\mu \gg \nu$ to signify that $\mu - \nu \in D(G)$. This being so we recall (cf. [**D***], p. 264), that there is a canonical procedure to construct, given $\mu \in D(G)$, a unitary representation π_μ of G. We have, if $\mu \gg \nu$, $\ker(\pi_\nu) \supseteq \ker(\pi_\mu)$. Below, given μ in $D(G)$, we shall write $\ker(\mu)$ in place of $\ker(\pi_\mu)$.

Let φ be a positive definite function given rise to by T. We denote by φ' the positive definite measure on G, obtained by extending $\varphi|K$ to G such that it be identically zero outside of K. We claim that $\ker(\varphi') = I$. In fact, $\varphi|K$ corresponds to a subrepresentation of $T|K$, and hence φ' to a subrepresentation of $T' = \mathrm{ind}_{K\uparrow G}(T|K)$ (cf. [**Bl**]). But, since T' is a factor representation, the kernel of the latter is again I. In this manner $\ker(\varphi') = I$.

Let dg, dk, and da be elements of the Haar measure resp. on G, K, and A such that $dg = dk \cdot da$. Given a continuous function f of compact support on G,

$$s \mapsto \int_K f(ks)\varphi(ks) \cdot dk$$

gives rise to one of the same kind on $A = G/K$. Hence, given μ in $M(A)$, there is $L(\mu) \in M(G)$ such that

$$L(\mu)(f) = \int_A \left(\int_K f(ks)\varphi(ks) \cdot dk \right) d\mu(a)$$

for any f of the sort considered above. Then we have $L(\delta_e) = \varphi'$ and also $L(D(A)) \subseteqq D(G)$. In fact, to see this, it is enough to check the special case, when $d\mu(a) = \chi(a) \cdot da$, χ being a character of A. But this is at once settled by noting that here

$$L(\mu)(f) = \int_G f(G) \cdot \varphi(G) \cdot \chi(\pi(G)) \cdot dg.$$

Let $d\chi$ be the Haar measure on $\widehat{A}$, and $(U_n) \subset \widehat{A}$ a decreasing sequence of open neighborhoods with compact support of the unity, the intersection of which is again the unity. If ψ_n is nonnegative, continuous, and of support in U_n such that $\int_{\widehat{A}} \psi_n(\chi) \cdot d\chi \equiv 1$, then putting

$$\mu_n(a) \equiv \int_{\widehat{A}} \psi(\chi)\langle a, \chi\rangle \cdot d\chi,$$

μ_n is positive definite, tending to one on each compact subset of A. If $\gamma_n > 0$ is such that $0 \leq \psi_n \leq \gamma_n$, then $\gamma_n \delta_e - \mu_n$ is positive definite on A. Hence, the same is true for $L(\gamma_n \delta_e - \mu_n) = \gamma_n \varphi' - L(\mu_n)$. In this fashion $L(\mu_n) \ll \gamma_n \varphi'$ and $I = \ker(\varphi') \subseteqq \ker(L(\mu_n))$. We have $L(\mu_n) = \varphi(\mu_n \circ \pi)$ which tends to φ uniformly on each compact subset of G. Hence we obtain finally

$$J = \mathrm{Ker}(\varphi) \supseteqq \bigcap_{n=1}^{\infty} \ker(L(\mu_n)) \supseteqq I$$

or $J \supseteqq I$, as claimed at the beginning.

2.5. Surjectivity; second step

In this section we shall assume that Γ is a central extension by a 1-torus of a free abelian group $\mathcal{Z}$. If T is a factor representation of Γ, then $T|\Gamma^c$ is the multiple of a character χ of Γ^c.

Let us add here that χ depends only on the kernel of T. In fact, if T_1 and T_2 are factor representations of Γ such that $\ker(T_1) = \ker(T_2)$, giving rise to the characters χ_1 and χ_2, then χ_j is weakly equivalent to $T_j|\Gamma^c$ $(j = 1, 2)$ and thus $\chi_1 = \chi_2$. We conclude from this that there is a map $f : \mathrm{Prim}(\Gamma) \to \mathrm{X}(\Gamma^c)$ such that if T is a factor representation of Γ satisfying $\ker(T) = J \in \mathrm{Prim}(\Gamma)$, then $T|\Gamma^c$ is a multiple of $f(J)$.

This being so, we have the following

PROPOSITION. *Let J be an element of* $\mathrm{Prim}_1(\Gamma)$ *and* $\chi = f(J)$. *Then* $J(\chi) \supseteqq J$.

In fact, we start by recalling (cf. [**Grl**]) that $\mathcal{Z}$ carries an invariant mean I. This is a linear form on the set of all complex-valued bounded functions on $\mathcal{Z}$, which is invariant under translation, satisfies $I(f) \geq 0$ if $f \geq 0$ and $I(1) = 1$. Let φ be a fixed, positive definite function corresponding to J on Γ. If f is an element of $C^*(\Gamma)$, then $(f, x\varphi)$ depends only on the class of x modulo the canonical central torus $\mathcal{T}$, and thus we can form $I_x((f, x\varphi)), = L(f)$ (say). The map $f \mapsto L(f)$ is continuous on $C^*(\Gamma)$ and hence also on $L^1(\Gamma)$. Hence there is an element ψ of $L^\infty(\Gamma)$ such that $L(f) = (f, \psi)$ $(f \in L^1(\Gamma))$. Clearly, ψ is positive definite.

LEMMA 1. *With the above notation, we claim that ψ is not zero.*

In fact, define f in $L^1(\Gamma)$ such that $f(u) \equiv \overline{u}$ $(u \in \mathcal{T})$ and let f be identically zero outside of $\mathcal{T}$. Since evidently $\varphi(u) \equiv u$ $(u \in \mathcal{T})$, we have that $(f, x\varphi) \equiv 1$, and thus $L(f) = 1$. Hence $\psi \neq 0$.

LEMMA 2. *For any element a of Γ, we have $a\psi = \psi$.*

In fact, let f be some element of $L^1(\Gamma)$. Then we have

$$(f, a\psi) = (a^{-1}f, \psi) = I_x((a^{-1}f, x\varphi)) = I_x((f, ax\varphi)).$$

Since I is $\mathcal{Z}$-invariant, the right-hand side is equal to $I_x((f, x\varphi)) = (f, \psi)$ and hence $(f, a\psi) \equiv (f, \psi)$ $(f \in L^1(\Gamma))$ and $a\psi = \psi$.

Given a continuous positive definite function φ on Γ, L^φ will denote the corresponding unitary representation of Γ.

LEMMA 3. *With the above notation, we have* $\ker(L^\psi) \supseteqq J$.

In fact,

a) Assume that f is some element of J. We claim that $(f, \psi) = 0$. Since $(f, \psi) = I_x((f, x\varphi))$, it suffices to show that $(f, x\varphi) \equiv 0$. Let U be a unitary representation of Γ and let $\xi \in H(U)$ be such that $\varphi(x) \equiv (U(x)\xi, \xi)$. Then $x\varphi(t) = (U(x^{-1})U(t)U(x)\xi, \xi)$, and thus

$$(f, x\varphi) = (U(x^{-1})U(f)U(x)\xi, \xi) = 0 \quad \text{by } U(f) = 0.$$

b) We claim that if $f \in J$ and $a \in \Gamma$, then $(f, a\psi) = 0$. To see this, it is enough to recall that $a\psi = \psi$.

c) If V is a unitary representation, and $\eta \in H(V)$ is such that $\psi(x) \equiv (V(x)\eta, \eta)$, then again $(a\psi)(x) \equiv (V(a^{-1})V(x)V(a)\eta, \eta)$. If f is in J, then so is f^*f, and therefore by b)

$$0 = (f^* \cdot f, a\psi) = \|V(f)V(a)\eta\|^2.$$

Since we can assume η to be cyclic, this implies that $V(f) = 0$; therefore, $f \in \ker(L^\psi)$ and $J \subseteqq \ker(L^\psi)$.

LEMMA 4. *Assume that $\chi = f(J)$ and that a is fixed in Γ. Then we have $\psi(a\gamma) = \chi(\gamma)\psi(a)$ for all $\gamma \in \Gamma^c$.*

In fact, let T be an irreducible representation of Γ such that $\ker(T) = J$. By Lemma 3 we have $\ker(L^\psi) \supseteqq \ker(T)$. In this manner L^ψ is weakly contained in T, and thus $L^\psi|\Gamma^c$ in $T|\Gamma^c$. In this manner, if $\chi = f(J)$, then $L^\psi|\Gamma^c = \chi \cdot I$. Assume again that $\psi(a) \equiv (V(a)\eta, \eta)$ $(V = L^\psi)$. If γ is in Γ^c, then

$$\psi(a\gamma) = (V(a)V(\gamma)\eta, \eta) = \chi(\gamma)\psi(a)$$

and therefore $\psi(a\gamma) \equiv \chi(\gamma)\psi(a)$ $(\gamma \in \Gamma^c)$.

LEMMA 5. *ψ is identically zero outside of* Γ^c.

In fact, if a is not in Γ^c, then there is an element γ in Γ such that $\gamma^{-1}a\gamma = ua$ $(u \neq 1)$. By Lemmas 2 and 4 we obtain in this fashion that

$$\psi(a) = \psi(\gamma^{-1}a\gamma) = u\psi(a).$$

Hence, since $u \neq 1$, $\psi(a) = 0$.

We use once more Blattner's theorem: If H is closed in G, μ positive definite on H, ν the extension of μ to G, which is zero outside of H, then $L^\nu = \mathrm{ind}_{H\uparrow G} L^\mu$. Let us apply this to the case when $G = \Gamma, H = \Gamma^c$, and $\mu = \chi$. Then, by what we saw above, L^ψ is unitarily equivalent to $V(\chi)$ and therefore, by Lemma 3,

$$J(\chi) = \ker(V(\chi)) = \ker(L^\psi) \supseteqq J.$$

Summing up, we have completed proving our Proposition. If J is in $\mathrm{Prim}_1(\Gamma)$, then there is a χ in $\mathcal{F}$ such that $J(\chi) \supseteqq J$.

For later use we note here the following.

LEMMA 6. *Let us assume that Γ is a central extension of a free abelian group by a 1-torus. If J is in $\mathrm{Prim}_1(\Gamma)$, then there is χ in $\mathcal{F}$ such that $J = J(\chi)$.*

In fact, by the Proposition above, there is $\chi_1 \in \mathcal{F}$ such that $J(\chi_1) \supseteqq J$. On the other hand, by the Proposition of 2.4, there is χ_2 in $\mathcal{F}$ satisfying $J \supseteqq J(\chi_2)$. In this fashion $J(\chi_1) \supseteqq J(\chi_2)$ and hence, by Lemma 5 in 2.3, $J(\chi_1) = J = J(\chi_2)$.

A similar result is valid if "class one" is replaced by "class minus one".

2.6. Proof of Lemma 4

The principal objective of this section is Lemma 4. Below, given a group G, we write $P(G)$ for the set of all normalized positive definite functions on G.

LEMMA 1. *Assume that G is separable and locally compact. Let J be a closed, 2-sided ideal in $C^*(G)$. Given φ in $P(G)$, we consider the following two conditions on J:*

(1) *$J \subseteqq \ker(L^\varphi)$,*

(2) *φ is orthogonal to J,*

and claim that they are equivalent.

a) Let us prove first that (1) implies (2). In fact, if $\varphi(g) \equiv (L^\varphi(g)\xi, \xi)$ then, for $a \in C^*(G)$, we have $(a, \varphi) = (L^\varphi(a)\xi, \xi)$, and hence $(a, \varphi) = 0$, if a is in $\ker(L^\varphi)$.

b) Assuming that φ is orthogonal to J, we want to show that $J \subseteqq L^\varphi$. If $a \in J$ and $b \in C^*(G)$, then we have $b^* \cdot a^* \cdot ab \in J$. Therefore, by virtue of our assumption

$$0 = (b^* \cdot a^* \cdot ab, \varphi) = \|L^\varphi(a)L^\varphi(b)\xi\|^2.$$

Hence, since ξ is cyclic, $L^\varphi(a) = 0$, proving our lemma.

Given a closed, two-sided ideal J in $C^*(G)$, we put $P_J = \{\varphi; \varphi \in P(G), \ker(\varphi) \supseteqq J\}$. Given a convex set S in a linear space, we write ES for the collection of all extremal elements in S.

LEMMA 2. *Assume that F is an open and invariant subgroup in G, and that J is a proper 2-sided closed ideal in $C^*(G)$. We put $I = J|F$. Let η be an element of EP_I. Then there is an ω in EP_J such that $\eta = \omega|F$.*

Let T be an irreducible representation of G such that $\ker(T) = J$.

a) By virtue of our assumption, L^η is weakly contained in $T|F$. Therefore (cf. [**D***], 18.1.4, Proposition) there is a sequence (η_n) of sums of positive definite functions, affiliated with $T|F$, such that η_n tends to η uniformly on each compact subset of F. Clearly, for $n = 1, 2, \ldots$ there is ω_n in P_J such that $\eta_n = \omega_n|F$. Since (ω_n) is bounded, passing, if necessary, to a subsequence we can assume that ω_n tends to ω weakly. Since F is open in G, $\omega_n|F$ tends weakly to $\omega|F$, and hence $\omega|F = \eta$. Since $\ker(\omega_n) \supseteqq \ker(T) = J$, by Lemma 1, ω_n is orthogonal to J, and thus so is ω. Therefore, by Lemma 1, $\ker(\omega) \supseteqq J$ and thus $\omega \in P_J$.

b) In this manner it remains to be shown that if $\eta \in EP_I$, then ω can be chosen in EP_J. In fact, let us put $C = \{\zeta; \zeta \in P_J, \zeta|F = \eta\}$. By a), C is nonempty. Evidently C is convex and weakly compact. Therefore we can conclude that EC, too, is nonempty. Assume that $\omega \in EC$ and that $\omega = \alpha\omega_1 + (1-\alpha)\omega_2$, with ω_1, ω_2 in P_J and $0 < \alpha < 1$. Then we have also

$$\eta = \omega|F = \alpha(\omega_1|F) + (1-\alpha)(\omega_2|F).$$

Since $\omega_j|F \in P_I$, we have $\omega_j|F = \eta$ and thus $\omega_j \in C$ $(j = 1, 2)$.

Summing up, by $\omega \in EC$ we have $\omega_1 = \omega = \omega_2$.

LEMMA 3. *Assume that F is a closed invariant subgroup in G. Let ω be an element of $P(G)$, such that its restriction η to F is in $EP(F)$. We claim that there is a countable subset A of G such that*

$$L^\omega|F = \bigoplus_{a \in A} a(L^\eta)$$

in the sense of unitary equivalence.

In fact, let us write $H = H(L^\omega), H_0 = H(L^\eta)$, $V = L^\omega|F$, and $W = L^\eta$. Our assumption implies that we can suppose $H \supseteqq H_0$ and that W is the component of V in H_0. For a fixed in G, let us put $H_a = L^\omega(a)H_0$. Then, since F is invariant in G, H_a is stable with respect to V, and the component of V in H_a is equal to $a(W)$. Also, evidently, H is spanned by $(H_a; a \in G)$. In this manner it will be enough to show that if H_1 is V-invariant in H, then there is a V-invariant subspace H_2 of H_1 such that $V|H_2$ is unitarily equivalent to cW for some $c \in G$. Let P_1 and P_c be the projections onto H_1 and H_c respectively. We can assume that P_1P_c is nonzero. Let us consider the polar decomposition $P_1P_c = U|P_1 \cdot P_c|$. We have $U^* \cdot U \le P_c$, U^*U commutes with V and $V|H_c$ is unitarily equivalent to $c(L^\eta)$. Hence $U^* \cdot U = P_c$ and it suffices to take $H_2 = U(H_c)$.

LEMMA 4. *Let J be some element of* $\mathrm{Prim}_1(\Gamma)$. *We claim that there is a factor representation V of Γ such that $J \subseteqq \ker(V)$ and $V|U^{\mathrm{c}}$ is transitive (in the sense of the Γ action of Γ in $\widehat{U}^{\mathrm{c}}$).*

In fact, let us put $I = J|U$. Then, by Lemma 2, we can select elements $\eta \in EP_I$ and $\omega \in EP_J$ such that $\omega|U = \eta$. Let us define $V = L^\omega$; then V is an irreducible representation of Γ and $J \subseteqq \ker(V)$. By Lemma 3, we can write

$$L^\omega|U = \bigoplus_{a \in A} a(L^\eta)$$

in the sense of unitary equivalence. L^η is a multiple of a character χ in $\mathcal{F}$. Hence $V|U^c$ is unitarily equivalent to

$$\bigoplus_{a\in A} a\chi$$

proving that $V|U^c$ is transitive.

2.7. Surjectivity; last step

Our objective below will be the following

LEMMA. *Assume that J is given in* $\mathrm{Prim}_1(\Gamma)$. *We suppose also that V is a factor representation of Γ such that* $\ker(V) = J$ *and $V|U^c$ is transitive. We claim that there is an element χ of $\mathcal{F}$ such that $J = J(\chi)$.*

In fact,

a) We start by recalling (cf. 2.2, Lemma 1) that given φ in $\mathcal{F}$, its stabilizer in Γ is equal to U. Hence (cf. [**AM**], Theorem 6, p. 58) there is a factor representation T of U such that

$$V = \mathrm{ind}_{U\uparrow\Gamma}\, T.$$

b) Below we shall make use of the following result due to Fell (cf. [**F1**], Theorem 4.1, p. 257). Assume that G is separable, locally compact, and K is a closed subgroup of G. Given unitary representations T_1 and T_2 such that $\ker(T_1) = \ker(T_2)$ of K, we have

$$\ker\Big(\mathrm{ind}_{K\uparrow G}\, T_1\Big) = \ker\Big(\mathrm{ind}_{K\uparrow G}\, T_2\Big).$$

We conclude from this that we can suppose from the beginning that $\Gamma = U$. In fact, assume that $\chi \in \mathcal{F}$ is such that

$$\ker(T) = \ker\Big(\mathrm{ind}_{U^c\uparrow U}\, \chi\Big).$$

We have then

$$\begin{aligned}\ker(V) = \ker\Big(\mathrm{ind}_{U\uparrow\Gamma}\, T\Big) &= \ker\Big(\mathrm{ind}_{U\uparrow\Gamma}\, \mathrm{ind}_{U^c\uparrow U}\, \chi\Big)\\ &= \ker\Big(\mathrm{ind}_{U^c\uparrow\Gamma}\, \chi\Big) = J(\chi)\end{aligned}$$

and thus $J = \ker(V) = J(\chi)$.

c) We recall that now $\Gamma = U$. Suppose that $V|\Gamma^c$ is a multiple of $\chi \in \mathcal{F}$. We select an irreducible unitary representation π of Γ_0 such that $\pi|U^c$ is a multiple of χ. We have clearly $U_\pi = U$. By [**AM**], Theorem 7, p. 61, if π^e is the projective extension of π to U, U^e is the corresponding group extension and

$$\Phi : U^e \to U^e/\Gamma_0 = M$$

then there is a factor representation ζ of class minus one of M such that $T = (\pi^e \otimes (\zeta \circ \Phi)^\sim)$. (Note that $\pi^e \otimes (\zeta \circ \Phi)$ is trivial on the extension torus; $(\ )^\sim$ indicates the corresponding representation of U.)

d) 1) Below we use another result due to Fell (cf. [**F2**], Theorem 2, p. 505). Let G be a separable locally compact group, and let S, T_1, T_2 be unitary representations of G such that $\ker(T_1) = \ker(T_2)$. Then

$$\ker(S \otimes T_1) = \ker(S \otimes T_2).$$

Since M is a central extension, by a 1-torus, of $U(\Gamma_0)$ that is free abelian, by Lemma 6 in 2.5 there is a character λ, of class minus one, of M^c such that

$$\ker(\zeta) = \ker\Big(\operatorname{ind}_{M^c \uparrow M} \lambda\Big).$$

We put $\Lambda = (\operatorname{ind}_{M^c \uparrow M} \lambda) \circ \Phi$. By the result just quoted of Fell, we conclude that

$$\ker(\pi^e \otimes (\zeta \circ \Phi)) = \ker(\pi^e \otimes \Lambda).$$

2) By Lemma 8 in 2.2, there are representations W and ρ of U^e such that W is quasi-equivalent to ρ, and ρ, when restricted to Γ_0, is unitarily equivalent to π. In this manner $\pi^e \otimes \Lambda$ is quasi-equivalent to $W \otimes \Lambda$.

e) By the end of the proof of Proposition 2.2, there is a character φ of U^c such that, when lifted to $(U^c)^e$ (same notation), we have

$$W \otimes \Lambda = \operatorname{ind}_{(U^0)^e \uparrow U^e} \varphi$$

and thus $\ker(T) = J(\varphi)$. Since $T|U^c$ is a multiple of χ, we have $\varphi \equiv \chi$ and thus $J = J(\chi)$, completing the proof of our lemma.

2.8. Summary

We sum up the considerations of 2.3–2.7 in the following.

PROPOSITION. *With the previous notation, the map $\chi \mapsto J(\chi)$ is continuous surjective from $\mathcal{F}$ onto* $\mathrm{Prim}_1(\Gamma)$; *we have $J(\chi_1) = J(\chi_2)$ if and only if the closures of $\Gamma\chi_j$ $(j = 1, 2)$ in $\mathcal{F}$ coincide.*

In fact,

a) For the continuity, cf. 2.3, Lemma 3.

b) We deal next with the surjectivity. Let $J \in \mathrm{Prim}_1(\Gamma)$ be given. By 2.6, there is a factor representation V of Γ such that $\ker(V) \supseteqq J$ and $V|U^c$ is transitive. Let us put $J_1 = \ker(V) \in \mathrm{Prim}_1(\Gamma)$. By 2.7 there is χ_1 in $\mathcal{F}$ such that $J_1 = J(\chi_2)$ and by 2.4, there is $\chi_2 \in \mathcal{F}$ such that $J \supseteqq J(\chi_2)$. In this manner we have $J(\chi_1) \supseteqq J(\chi_2)$ and thus by 2.3, $J(\chi_1) = J = J(\chi_2)$. Thus $J : \mathcal{F} \to \mathrm{Prim}_1(\Gamma)$ is surjective.

c) To complete our proof it is enough to recall that, by 2.3, $J(\chi_1) = J(\chi_2)$ if and only if the closures of $\Gamma\chi_j$ $(j = 1, 2)$ coincide.

CHAPTER III

Existence of Characters

We start by recalling that, given a separable, locally compact group G, a unitary factor representation V of G is called normal if the von Neumann algebra it generates is semifinite, such that there is an element of $C^*(G)$ the image of which is nonzero and traceable. The objective of this chapter is to show that if G is a connected and simply connected Lie group and J a primitive ideal of $C^*(G)$, then there is a normal factor representation V of G such that $\ker(V) = J$ and that, up to quasi-equivalence, V is uniquely determined by these properties. Below we present our original proof of this statement (cf. [**Lp 5**], Theorem 1, p. 119). By now this can be derived from a more general proposition due to D. Poguntke (cf. 3.9 below). However, components of our proof will play a substantial role in the next chapter.

3.1. Some subgroups of G

Let L be the first derived group of G. In this first section we shall establish the following. Given an element π of $\widehat{L}$, there is a canonical procedure assigning to π a closed subgroup K_π, containing L, of G with the following property. There is an irreducible representation ρ of K_π, which on L restricts to π, and for which $\mathrm{ind}_{K_\pi \uparrow G}\, \rho$ is a semifinite factor representation.

Conversely, any irreducible representation σ of K_π such that $\sigma|L = \pi$ gives rise, through induction, to a semifinite factor representation.

LEMMA 1. *L is of type* I.

In fact, let $\mathfrak{G}$ be the Lie algebra of G. By the theorem of Ado, we can assume that it is presented as a subalgebra of the Lie algebra of all endomorphisms of a finite-dimensional real vector space. Then if $I = [\mathfrak{G}, \mathfrak{G}]$, I is an algebraic Lie algebra (cf. [**Ch II**], Théorème 15, p. 177). Hence by what we saw in 1.2, L is of type I.

Let π be a fixed element of $\widehat{L}$. We note that its stabilizer G_π in G is closed. In fact, since $L = (I)$, $\widehat{L}$ is standard. Thus it is enough to note (cf. [**AM**], p. 16) that if E is a countably separated Borel space, acted upon by a separable, locally compact group G such that for any x in E, $a \mapsto ax$ is Borel, then G_x is closed.

We recall (cf. [**AM**], Proposition 10.3, p. 59) that π admits a projective extension π^e to G_π. Thus, in particular, there is a complex-valued, measurable function $\omega : G_\pi \times G_\pi \to \mathbb{T}$ such that we have for any $a, b \in G_\pi$

$$\pi^e(a)\pi^e(b) = \omega(a,b)\pi^e(ab).$$

By loc. cit. we can assume

$$\omega(al_1, bl_2) = \omega(a,b) \qquad (l_1, l_2 \in L, a, b \in G_\pi)$$

and that $\omega(a, a^{-1}) = 1$. By our hypothesis, G_π/L is isomorphic to $\mathbb{R}^a \times \mathbb{T}^b$. Hence (cf. [**AM**], p. 188) we are permitted to suppose that ω is the inflation from $G_\pi/L \times$

G_π/L of a skew-symmetric, bilinear form with values in $\mathbb{T}$. Below we shall assume that such a choice has been fixed. We define, as usual, G_π^e as

$$\{(a,u); a\in G_\pi, u\in T \text{ such that } (a,u)(b,v)=(ab,\omega(a,b)uv)\}.$$

This is now a Lie group. We note next that $\{(l,e); l\in L\}$ is a closed, invariant subgroup of G_π^e.

In fact, we have $(a,u)^{-1}=(a^{-1},u^{-1})$. Also, if $l\in L, a\in G_\pi$ and $u\in T$, then

$$\begin{aligned}(a,u)(l,1),(a,u)^{-1} &= (a,u)(l,e)(a^{-1},u^{-1})\\ &= (a,u)(la^{-1},\omega(l,a^{-1})u^{-1}) = (ala^{-1},\omega(a,la^{-1})) = (ala^{-1},e)\end{aligned}$$

and hence $(a,u)(l,e)(a,u)^{-1}=(ala^{-1},e)$, proving our statement. Whenever convenient, we shall in what follows write l in place of $(l,1)$ $(l\in L)$. We form the Mackey group M_π by

$$M_\pi = G_\pi^e/L \quad (\leftarrow G_\pi^e:\phi).$$

LEMMA 2. *Assume that A is a closed subgroup, containing L, of G. Let ρ be a unitary representation of A that, on L, restricts to π. We claim that A is a subgroup of G_π.*

In fact, let a be an element of A. We have then

$$a\pi = a(\rho|L) = a\rho|L = \rho|L = \pi$$

which implies $a\pi=\pi$, or that a is in G_π.

Given a subset S of G_π, if convenient, we shall write S^e for the subset

$$\{(s,u); s\in S, u\in\mathbb{T}\}\subset G_\pi^e.$$

LEMMA 3. *Let A be a closed subgroup of G such that $L\subseteqq A\subseteqq G_\pi$. Then the following conditions are equivalent*:

(1) *There is a representation ρ of A extending π.*

(2) *$\phi(A^e)$ is abelian in M_π.*

In fact

$$(1)\Rightarrow(2).$$

Our assumption implies that there is a map $f: A\to\mathbb{T}$ such that $\pi^e(a)=f(a)\rho(a)$.

a) We conclude from this that for any $a,b\in A$ we have

$$\omega(a,b)=f(a)f(b)/f(ab).$$

In fact, since $\pi^e(a)\pi^e(b)=\omega(a,b)\pi^e(ab)$, we conclude that $f(a)f(b)\rho(a)\rho(b)=\omega(a,b)f(ab)\rho(ab)$ or that $\omega(a,b)f(ab)=f(a)f(b)$.

b) We have for $a\in A$ and $l\in L$, $f(al)=f(a)$. In fact, $\pi^e(a)\pi^e(l)=\omega(a,l)\pi^e(al)$ and $\omega(a,l)=1$.

c) Since G/L is abelian, we have $ab=bal$ for some $l\in L$, whence, by b) $f(ab)=f(ba)$ and thus, by a) $\omega(a,b)=\omega(b,a)$.

d) With notation as above, we get:

$$\begin{aligned}(a,u)(b,v) &= (ab,\omega(a,b)uv) = (ba,\omega(b,a)uv)(l,e)\\ &= (b,a)(a,u)(l,e).\end{aligned}$$

Hence, if $L \subsetneqq A \subsetneqq G_\pi$ and π extends to A, then $\phi(A^e)$ is abelian in M_π or (1) $\Rightarrow$ (2).

(2) $\Rightarrow$ (1)

Here we assume that $\phi(A^e)$ is abelian in M_π.

a) We form the quotient $G_\pi/L = H$ and denote by ψ the canonical morphism from G_π into H. Let β be the cocycle on $H \times H$, which when lifted to $G_\pi \times G_\pi$, coincides with ω. We can assume

$$M_\pi = \{(c,u); c \in H, u \in T \text{ such that } (c,u)(d,v) = (cd, \beta(c,d)uv)\}.$$

We have also $\phi((a,u)) = (\psi(a), u)$ $(a \in G_\pi, u \in \mathbb{T})$.

b) We write $\mathcal{T} = \{(e,u); u \in \mathbb{T}\} \subset M_\pi$. Since $\phi(A^e)$ is abelian, there is a closed subgroup B in $\phi(A^e)$ such that $\phi(A^e) = B \times \mathcal{T}$ (direct product of groups).

We write τ for the projection onto the first component. We claim that there is a map $g : \psi(A) \to \mathbb{T}$ such that for any $b \in \psi(A)$, (b,u) lies in B if and only if $u = g(b)$. In fact, if (b,u) and (b,v) are in B, then (e, uv^{-1}) lies in $B \cap \mathcal{T}$ and hence $u = v$.

c) We claim that if $a, b \in \psi(A)$, then $\beta(a,b) = g(ab)/(g(a)g(b))$.

In fact,

1) We have $\tau((a,1)(b,1)) = \tau((ab, \beta(ab))) = \tau((ab,1)(e,\beta(ab))) = \tau((ab,1)) = (ab, g(ab))$.

We conclude from this that $\tau((a,1)(b,1)) = (ab, g(ab))$.

2) We have

$$\tau((a,1)(b,1)) = \tau((a,1))\tau((b,1)) = (a, g(a))(b, g(b)) = (ab, \beta(a,b)g(a)g(b)).$$

This being so, 1) and 2) above jointly imply that for $a, b \in \psi(A)$

$$\beta(a,b) = g(ab)/(g(a)g(b))$$

as claimed above.

We write $f(a) \equiv 1/g(\psi(a))$ $(a \in A)$ and conclude that $\omega(a,b) = f(a)f(b)/f(ab)$.

d) We set $\rho(a) = \pi^e(a)/f(a)$. To complete the proof that (2) implies (1) it is enough to show that ρ is an extension of π to A.

But this is so, since if $a, b \in A$, then

$$\begin{aligned}\rho(a)\rho(b) &= \pi^e(a)\pi^e(b)/(f(a)f(b)) \\ &= \pi^e(ab)/f(ab) = \rho(ab).\end{aligned}$$

Summing up, we have proved that if A is closed in G such that $L \subsetneqq A \subsetneqq G_\pi$, then π extends to A if and only if $\phi(A^e)$ is abelian in M_π, which is the claim of Lemma 3.

Given a group G and a subset A of G we shall write

$$C(A) = \{g; g \in G \text{ such that } ga = ag \text{ for all } a \in A\}.$$

LEMMA 4. *Assume that $\pi \in \widehat{L}$ is fixed, and A is a closed subgroup of G such that $L \subsetneqq A \subsetneqq G_\pi$ and that ρ extends π to A. We claim that*

$$G_\rho = \phi^{-1}(C(\phi(A^e)))/\mathbb{T}.$$

In fact, let us show first that the right-hand side is contained in the left-hand side. Let c be an element of the right-hand side. Let B be the closed subgroup of G_π, generated by c and A. Then $\phi(B^e)$ is abelian and thus, by Lemma 3, there is a representation σ of B, extending π.

Since ρ extends π to A, there is a character φ of $\mathrm{X}(A)$, identically one on L and such that $\sigma|A = \varphi\rho$. Therefore $\varphi c\rho = \varphi\rho$, whence $c\rho = \rho$ and thus c belongs to G_ρ (if $\phi((c,1))$ commutes with $\phi(A^e)$). In this fashion we have shown that the right-hand side is contained in G_ρ. We proceed now to establish the opposite implication.

a) We start by showing that if c is a fixed element of G_ρ, then there is a unitary operator V on the space of π depending on c only, such that

$$(c\pi^e)(a) = V\pi^e(a)V^{-1}$$

for all a in A. To this end:

1) We denote by f the map of A into $\mathbb{T}$ such that $\pi^e = f\rho$ on A. As we saw in the proof of Lemma 3, we have for any $b \in A$ and $l \in L : f(bl) \equiv f(b)$.

2) Since $c \in G_\rho$, there is a unitary operator V, depending on c only, such that

$$\rho(c^{-1}ac) = V\rho(a)V^{-1} \quad (a \in G).$$

From all this we conclude that if $a \in A$ then

$$(c\pi^e)(a) = \pi^e(c^{-1}ac) = f(c^{-1}ac)\rho(c^{-1}ac).$$

But $c^{-1}ac = cl$ for some $l \in L$, and thus $f(c^{-1}ac) = f(al) = f(a)$, whence

$$(c\pi^e)(a) = f(a)V\rho(a)V^{-1} = V(\pi^e(a))V^{-1} \quad (a \in A)$$

or

$$(c\pi^e)(a) \equiv V(\pi^e(a))V^{-1}.$$

b) Let b, c be elements of G_π, c fixed. We claim that

$$(c\pi^e)(b) = \eta(b)(\pi^e(c))^{-1} \cdot \pi^e(b) \cdot \pi^e(c)$$

where

$$\eta(b) = \overline{\omega(c^{-1}b, c)} \cdot \overline{\omega(c^{-1}, b)}.$$

In fact, we have

$$\begin{aligned}(c\pi^e)(b) &= \pi^e(c^{-1} \cdot bc) = \pi^e((c^{-1}b)c) \\ &= \overline{\omega(c^{-1}b, c)}\pi^e(c^{-1}b)\pi^e(c).\end{aligned}$$

But we have also $\pi^e(c^{-1}b) = \overline{\omega(c^{-1}, b)}\pi^e(c^{-1})\pi^e(b)$ and thus

$$(c\pi^e)(b) = \eta(b)\pi^e(c^{-1})\pi^e(b)\pi^e(c).$$

In this manner, to obtain the desired conclusion it is enough to note that $\pi^e(c^{-1}) = (\pi^e(c))^{-1}$.

c) We note next that $\eta(a) \equiv 1$ for all $a \in A$. In fact, a) implies that $(c\pi^e)(a) = V\pi^e(a)V^{-1}$. From (b) we can conclude that

$$(c\pi^e)(a) \equiv \eta(a)(\pi^e(c))^{-1} \cdot \pi^e(a) \cdot \pi^e(c) \quad (a \in A).$$

Therefore, there exists a unitary operator U, depending on c only, such that

$$\pi^e(a) \equiv \eta(a)U^{-1}\pi^e(a)U \quad (a \in A).$$

Since evidently $\eta(l) \equiv 1 (l \in L)$, hence $U = \omega I$ ($|\omega| = 1$).

d) We conclude from all this that by $\eta(a) = 1$, we have $1 = \omega(c^{-1}a, c)\omega(c^{-1}, a)$ ($a \in A$). This implies that for all a in A: $(c^{-1}ac, 1) = (c, 1)^{-1} \cdot (a, 1)(c, 1)$. In fact, since $(c, 1)^{-1} = (c^{-1}, 1)$, the right-hand side is equal to $(c^{-1}a, \omega(c^{-1}a))(c, 1) =$

$(c^{-1}ac, \omega(c^{-1}a, c)\omega(c^{-1}, a))$ which, by $\eta(a) = 1$, is the same as $(c^{-1}ac, 1)$. Since there is l in L such that $c^{-1}ac = al$, $\phi((c, 1))$ commutes with $\phi(A^e)$ if $c \in G_\rho$.

Summing up all, if π extends to the unitary representation ρ on A, then

$$G_\rho = \phi^{-1}(C(\phi(A^e)))/\mathbb{T}$$

completing the proof of Lemma 4.

Below we keep $\pi \in \widehat{L}$ fixed.

LEMMA 5. *Let us denote by $\mathcal{A}$ the totality of all those connected subgroups A of G such that $L \subseteqq A$ and π admits a $(G_\pi)_0$-invariant extension to A. We claim that $\mathcal{A}$ admits a well-defined maximal element.*

In fact, let us write $\Gamma = M_\pi$. Γ is a central extension by a 1-torus of a closed subgroup of $\mathbb{R}^n$. This being said, we shall use the notation of Chapter 2.

a) Let A be an element of $\mathcal{A}$. By our assumption and Lemma 2, we have $A \subseteqq G_\pi$ and thus we can form $\phi(A^e) \subseteqq \Gamma$. By Lemma 4, $\phi(A^e)$ commutes with Γ_0 and hence, since $\phi(A^e)$ is connected, we have $\phi(A^e) \subseteqq (\Gamma_0)^c$.

b) In this manner it suffices to show that if

$$\Pi \underset{\text{def}}{=} \phi^{-1}((\Gamma_0)^c)/\mathbb{T}$$

then $\Pi \in \mathcal{A}$. In fact, we note first that Π is connected in G_π. Since $\phi(A^e) = (\Gamma_0)^e$ is abelian, by Lemma 3 there is an extension ρ of π to Π. Since $\Gamma_0 \subset C(\phi(A^e))$, by Lemma 4 ρ is $(G_\pi)_0$-invariant. From this we conclude that Π is the maximal element called for in the statement of our lemma.

Given π and ρ as above, we shall write $\mathcal{U} = G_\rho$. If U is the centralizer of $(\Gamma_0)^c$ in Γ, then by Lemma 4, we have $\mathcal{U} = \phi^{-1}(U)/\mathbb{T}$.

LEMMA 6. *Let us write $\mathcal{B}$ for the collection of all those closed subgroups containing L to which π admits a $\mathcal{U}$-invariant extension. We claim that $\mathcal{B}$ admits a well-defined maximal element.*

In fact,

a) If A is in $\mathcal{B}$, then $\phi(A^e) \subseteqq U^c$. If ρ is an extension of π to A, then by Lemma 4 and the assumption above, we have

$$\mathcal{U} = \phi^{-1}(C(\phi(A^e)))/\mathbb{T}$$

and hence $\phi(\mathcal{U}^e)$ and $\phi(A^e)$ commute. Since $\phi(\mathcal{U}^e) = U$, we have $\phi(A^e) \subseteqq U^c$.

b) In this manner it suffices to show that if we define B by $\phi^{-1}(U^c)/\mathbb{T}$, then B is in $\mathcal{B}$. In fact, $\phi(B^c) = U^c$ is abelian and thus, by Lemma 3, π extends to B as ρ (say) and

$$G_\rho = \phi^{-1}(C(\phi(B^e))) = \phi^{-1}(U)/\mathbb{T} = \mathcal{U}.$$

From this we conclude that B is the maximal element, the existence of which was asserted in our lemma.

The following lemma is evident.

LEMMA 7. *Let us write K for the maximal element of Lemma 6. Let ρ be an extension of π to K. Then $G_\rho = \mathcal{U}$, and K_0 is the maximal element of $\mathcal{A}$ as in Lemma 5.*

Below, if the context requires us to indicate its dependence on π, we shall write K_π. We sum up the construction for given $\pi \in \widehat{L}$ of K_π as follows.

1) Take the maximal element among the connected subgroups A, $L \subseteqq A \subseteqq G_\pi$, to which π admits a $(G_\pi)_0$-invariant extension (cf. Lemma 5). Let Π be this element and ρ an extension of π to Π. We write $\mathcal{U} = G_\rho$, which is independent of ρ if $\rho|L = \pi$.

2) K_π is the maximal element in the collection of all those closed subgroups A, $L \subseteqq A \subseteqq G_\pi$, to which π admits a $\mathcal{U}$-invariant extension (cf. Lemma 6).

We have

$$K = \phi^{-1}(U^{\mathrm{c}})/\mathbb{T}, \qquad \mathcal{U} = \phi^{-1}(U)/\mathbb{T}.$$

The following lemma is evident.

LEMMA 8. *If φ is an automorphism of G and $\pi \in \widehat{L}$, then we have*

$$K_{\varphi\pi} = \varphi(K_\pi).$$

Unexplained notation as above.

LEMMA 9. *We claim that $\operatorname{ind}_{K\uparrow G} \rho$ is a semifinite factor representation. It is of type* I *if and only if $\mathcal{U}/(G_\pi)_0K$ is finite.*

In fact,

a) We recall from Chapter 2 the following facts. Let Z be a closed subgroup of $\mathbb{R}^n$. Assume that Γ satisfies

$$1 \to \mathbb{T} \xrightarrow{\text{central}} \Gamma \to Z \to 1.$$

Let U be the centralizer of $(\Gamma_0)^{\mathrm{c}}$. We write $\mathcal{T}$ for the canonical torus in Γ. We set $\mathcal{F} = \{\chi; \chi \in \mathrm{X}(U^{\mathrm{c}}), \chi|\mathcal{T} = \overline{\mathrm{Id}} : \mathcal{T} \to \mathcal{T}\}$. Given $\chi \in \mathcal{F}$ we define $V(\chi) = \operatorname{ind}_{U^{\mathrm{c}}\uparrow\Gamma} \chi$. All this being so we claim, in particular, the following.

1) $V(\chi)$ is a semifinite factor representation such that the restriction of $V(\chi)$ to $\mathcal{T}$ is quasi-equivalent to the conjugate of the identity map of $\mathcal{T}$ onto itself.

2) $V(\chi)$ is of type I if and only if $U/U^{\mathrm{c}} \cdot \Gamma_0$ is finite. We recall (cf. b) in the proof of Lemma 6) that K is defined as $\phi^{-1}(U^{\mathrm{c}})/\mathbb{T}$.

b) Assume that $\pi \in \widehat{L}$ is given, fixed. We have

$$V(\chi) \circ \phi = \operatorname{ind}_{K^{\mathrm{e}}\uparrow G^{\mathrm{e}}_\pi}(\chi \circ \phi) = \tau(\chi) \quad \text{(say)}.$$

(Note: Strictly speaking, $\tau(\chi)$ is a representation of G^{e}_π; since $\tau(\chi)|\mathbb{T} \equiv 1$, we identify it to the corresponding representation of $G^{\mathrm{e}}_\pi/\mathbb{T}$.)

We conclude from this that $\tau(\chi)$ is a factor representation. We recall that $\mathcal{U} = \phi^{-1}(U)/\mathbb{T}$, and clearly

$$U/\Gamma_0 \cdot U^{\mathrm{c}} = \mathcal{U}/(G_\pi)_0K$$

and thus $\tau(\chi)$ is of type I if and only if $\mathcal{U}/(G_\pi)_0K$ is finite.

c) Below we write π^{e} also for $\pi^{\mathrm{e}}|K^{\mathrm{e}}$. Let us put $\mu = \chi \circ \phi \in \mathrm{X}(K^{\mathrm{e}})$. We have

$$\pi^{\mathrm{e}} \otimes \tau(\chi) = \pi^{\mathrm{e}} \otimes \left(\operatorname{ind}_{K^{\mathrm{e}}\uparrow G^{\mathrm{e}}_\pi} \mu\right) = \operatorname{ind}_{K^{\mathrm{e}}\uparrow G^{\mathrm{e}}_\pi}(\pi^{\mathrm{e}}\mu).$$

We write σ for the representation corresponding to $\mu\pi^{\mathrm{e}}$ of $K = K^{\mathrm{e}}/\mathbb{T}$. If σ_1 stands for the representation corresponding to $\pi^{\mathrm{e}} \otimes \tau(\chi)$, of $G_\pi = G^{\mathrm{e}}_\pi/\mathbb{T}$, then we have clearly $\sigma|L = \pi$ and $\sigma_1|L \approx \pi$.

We have $T = \operatorname{ind}_{G_\pi\uparrow G} \sigma_1 = \operatorname{ind}_{K\uparrow G} \sigma$, and thus T is a semifinite factor representation of G. It is of type I if and only if $\mathcal{U}/(G_\pi)_0K$ is finite.

d) Let now ρ be any extension of π to K. Then there is a character φ in $\mathrm{X}(K)$ such that $\varphi|L \equiv 1$ and $\rho = \varphi\sigma$, and thus all the conclusions derived above for σ remain valid for ρ.

For $\pi \in \widehat{L}$ fixed, we put $\mathfrak{G}(\pi) = \{\rho; \rho \in \widehat{K}$ such that $\rho|L = \pi\}$.

LEMMA 10. *Let π_1, π_2 be elements of $\widehat{L}$ and $\rho_j \in \mathfrak{G}(\pi_j)$ such that $\mathrm{ind}_{K_{\pi_1}\uparrow G}\,\rho_1$ is quasi-equivalent to $\mathrm{ind}_{K_{\pi_2}\uparrow G}\,\rho_2$. Then $K_{\pi_1} = K_{\pi_2}$ and there is $a \in G$ such that $a\rho_1 = \rho_2$.*

In fact, (for the central decomposition to be used below, we refer to [**D***], 8.4, p. 148),

a) Let us put $T_j = \mathrm{ind}_{K_{\pi_j}\uparrow G}\,\rho_j$ $(j = 1,2)$. Since L is of type I, the central disintegration of $T_j|L$ is carried by $G\pi_j$ $(j = 1,2)$. If T_1 is quasi-equivalent to T_2, then so is $T_1|L$ to $T_2|L$ and thus $G\pi_1 = G\pi_2$. Hence there is an element a of G such that $\pi_2 = a\pi$ and thus, by Lemma 7, $K_{\pi_1} = K_{\pi_2}$. Below we shall write K for K_{π_j} $(j = 1,2)$.

b) Since K may fail to be of type I, in completing the proof of our lemma more care is needed. We start by recalling (cf. [**D***], 8.4.2, p. 148) that if T is a continuous unitary representation of K then there is a standard measure μ on $\widehat{K}$, and a field of μ-measurable factor representations $\{T(\rho)\}$ on $\widehat{K}$ such that

$$T = \int_{\widehat{K}}^{\oplus} T(\zeta)\, d\mu(\zeta).$$

Also, $(R(T))^{\mathrm{c}}$ corresponds to the family of diagonalizable operators; in other words, our decomposition is central. The measure μ, up to equivalence well-determined by T, will be referred to below, by abuse of language, as the central measure of T.

Our objective below will be to show that if $\pi \in \widehat{L}$, $\rho \in \mathfrak{G}(\pi)$ and $T = \mathrm{ind}_{K\uparrow G}\,\rho$, then the central measure of $T|K$ is carried by $G\rho$ and there is equal to the invariant measure obtained from G/G_ρ. If so, we can complete the proof of Lemma 10 very much as in the situation of a) above. If T_1 is quasi-equivalent to T_2, then so is $T_1|K$ to $T_2|K$. In this case, by [**D***], 8.4.4 (p. 151) the corresponding central measures are equal, whence, by the claim above, $G\rho_1 = G\rho_2$.

c) 1) Let us write $A = G/K$. If $f : A \to G$ is a Borel cross section and da the invariant measure on A, then evidently

$$T|K = \int_A^{\oplus} f(a)\rho\, da.$$

2) We set $B = \mathcal{U}/K$, $\Lambda = A/B$.

Suppose that φ is a Borel cross section $\Lambda \to A$. Next we make the following observations.

α) The map $(\lambda, b) \mapsto \varphi(\lambda) + b = \omega(\lambda, b)$ (say) is a Borel bijection $\Lambda \times B \to A$. We can assume that $d\lambda \cdot db$ corresponds to da.

β) We put $g(\lambda) \equiv f(\varphi(\lambda))$ $(\lambda \in \Lambda)$. Evidently, $g : \Lambda \to G$ is a Borel cross-section.

γ) $f(\omega(\lambda, b))$ is congruent, modulo K, to $g(\lambda)f(b)$. We conclude from all this that, in the sense of unitary equivalence, we obtain

$$T|K = \int_A^{\oplus} f(a)\rho \cdot da = \int_{\Lambda\times B}^{\oplus} f(\omega(\lambda, b))\rho \cdot d\lambda \cdot db = \int_{\Lambda\times B}^{\oplus} g(\lambda)f(b)\rho \cdot d\lambda\, db.$$

Put $U = \int_B f(b)\rho \cdot db$. Since $\mathcal{U} = G_\rho$ and $f(B) \subset \mathcal{U}$, U is quasi-equivalent to ρ.

Let us set $T(\lambda) = g(\lambda)U$ $(\lambda \in \Lambda)$. Our next objective is to show that the decomposition

$$\int_{\Lambda}^{\oplus} T(\lambda)\, d\lambda$$

is central. Evidently, we can assume that $U \equiv \rho$.

d) 1) Let us put $C = G_\pi/\mathcal{U}$. Writing $H = \Lambda/C$, we form a Borel cross-section $h : H \to \Lambda$. This being so, proceeding as in (2), with analogous notation we obtain

$$T|K = \int_{H\times C} h(\eta)g(c)\rho \cdot d\eta \cdot dc.$$

2) If $a \in G_\pi$, we can write $a\rho = \varphi_a\rho$ where φ_a is in $\mathrm{X}(K)$ such that $\varphi_a|L \equiv 1$. We note that $\varphi_a \equiv \varphi_b$ if and only if $a \equiv b$ $(\mathcal{U})$.

3) Let us write

$$T_1 = \bigoplus_{c\in C} \omega_c, \qquad T_2 = \int_H^{\oplus} h(\eta)\rho \cdot d\eta.$$

We denote by Z_j the ring of decomposition for T_j $(j = 1, 2)$ and by Z that for $T|K$. With obvious identifications we have

$$T|K = T_1 \otimes T_2, \qquad Z = Z_1 \otimes Z_2.$$

Hence, to complete our proof, it will be enough to establish that $Z_1 \otimes Z_2 = (R(T_1 \otimes T_2))^{\mathrm{c}}$.

e) 1) We have $Z_1 = R(T_1)$.

2) $R(T_2|L) = Z_2'$. In fact, since $L = (I)$, $\widehat{L}$ is standard; hence, for any π fixed, $\eta \mapsto h(\eta)\pi$ is a Borel isomorphism of H with its image in $\widehat{L}$. We recall (cf. [**D***], 4.6.2, p. 95) that there is a Borel map $\pi : \widehat{L} \to \mathrm{Irr}(L)$ such that $\pi(\zeta)$ is of the class of ζ $(\zeta \in \widehat{L})$. We write $d\nu(\zeta)$ for the image of $d\eta$ in $\widehat{L}$. We conclude from this that, up to unitary equivalence,

$$T_2|L = \int_{\widehat{L}}^{\oplus} \pi(\zeta) \cdot d\nu(\zeta)$$

whence (cf. [**D***], 8.6.4, p. 155) $R(T_2|L) = Z_2'$.

3) α) We have $I \otimes Z_2 \subset (R(T_1 \otimes T_2))^{\mathrm{c}}$.

β) Since the operators of T_2 are all decomposable, we have $R(T_2) = Z_2'$ whence

$$I \otimes T_2(k) \in R(T_1 \otimes T_2)$$

and $T_1(k) \otimes I \in R(T_1 \otimes T_2)$.

We have by 1): $(Z_1 \otimes I) \subset (R(T_1 \otimes T_2))^{\mathrm{c}}$ whence $Z_1 \otimes Z_2 \subset (R(T_1 \otimes T_2))^{\mathrm{c}}$.

4) We recall that $T|K$ is decomposed into a continuous direct sum of irreducibles, whence we conclude that $(R(T_1 \otimes T_2))^{\mathrm{c}} \subsetneqq Z \subsetneqq Z_1 \otimes Z_2$, and $Z_1 \otimes Z_2 = (R(T_1 \otimes T_2))^{\mathrm{c}}$ as claimed at the end of d).

We recall that what we are up to is a proof that $T|K$ is carried by $G\rho$ and there is equal to the invariant measure of G/G_ρ.

f) Consider the map $T(\lambda) : \Lambda \to \mathrm{Fac}(K)$. It is Borel and injective: if $\lambda_1 \neq \lambda_2$, the corresponding values cannot be quasi-equivalent.

We write S for the image of Λ in $\mathrm{Fac}(K)$. Since $\mathrm{Fac}(K)$ is standard, S is Borel. Next we recall the following result. Let p be the canonical projection $\mathrm{Fac}(K) \to \widehat{K}$. Then $p(S)$ is Borel and $p|S$ is an isomorphism $S \to p(S)$. But $p(S) = G_S$.

All this being so we conclude that the decomposition $T|K = \int_\Lambda^\oplus T(\lambda)\cdot d\lambda$ is central. This implies by transfer that $T|K = \int_{G/G_\rho} T(\zeta)\cdot d\mu(\zeta)$ (where class $T(\zeta) = \zeta \in \widehat{K}$) is central.

Final conclusion: The central measure of $T|K$ is carried by $G\rho$ and there coincides with the invariant measure of G/G_ρ.

3.2. The orbits of J

We continue to write $\mathfrak{G}$ for a real, finite-dimensional Lie algebra, and G for the corresponding connected and simply connected Lie group. We assume, as we can, that $\mathfrak{G}$ is presented as a Lie algebra of endomorphisms on a finite-dimensional real vector space. We denote by $\widetilde{\mathfrak{G}}$ the algebraic hull of $\mathfrak{G}$; that is, $\widetilde{\mathfrak{G}}$ is the smallest algebraic Lie algebra containing $\mathfrak{G}$. We recall (cf. [**Ch II**], Théorème 13, p. 173), that $[\widetilde{\mathfrak{G}}, \widetilde{\mathfrak{G}}] = [\mathfrak{G}, \mathfrak{G}]$. Let $\widetilde{G}$ be a connected and simply connected group belonging to $\widetilde{\mathfrak{G}}$.

Let us set $L = [G, G]$. As we saw in Lemma 1 of 3.1, L is a connected and simply connected Lie group, which is of type I.

Let M be a separable, locally compact group, and H a closed subgroup of M such that $[M, M] \subseteqq H$. We denote by $\mathcal{F}$ the annihilator of $[M, M]$ in $\mathrm{X}(H)$. We form the direct product $N = M \times \mathcal{F}$. Let dx be the element of the right-invariant Haar measure on H. For $b \in M$ we put $d(bxb^{-1}) = \Delta(b)\cdot dx$. We write $\mathfrak{H}$ for the group C^*-algebra $C^*(H)$ of H. We consider $\mathrm{Aut}(\mathfrak{H})$ as topologized with pointwise convergence.

LEMMA 1. *There is a continuous representation of N in* $\mathrm{Aut}(\mathfrak{H})$ *such that the action of $a = (b, \chi)$ on $f \in \mathcal{L}(H)$ is equal to*

$$x \mapsto \chi(x) f(b^{-1}xb)/\Delta(b) \quad (x \in H).$$

In fact, for $f \in \mathcal{L}(H)$ and $a \in M$ we set $(af)(x) = f(a^{-1}xa)/\Delta(a)$.

a) We observe that if π is a representation of H and $a \in M$, then we have $\pi(af) = (a^{-1}\pi)(f)$. We have, in fact,

$$\begin{aligned}\pi(af) &= \left(\int_H \pi(x) f(a^{-1}xa)\cdot dx\right)/\Delta(a)\\ &= \left(\int_H \pi(axa^{-1}) f(x)\cdot d(axa^{-1})\right)/\Delta(a)\\ &= \int_H \pi(xax^{-1}) f(x)\,dx = (a^{-1}\pi)(f).\end{aligned}$$

We conclude from this that if $\|\ \|$ is the norm on $C^*(H)$, then $\|af\| = \|f\|$ $(a \in M, f \in \mathfrak{H})$.

b) We note that if $f \in \mathcal{L}(H)$ and $\chi \in \mathrm{X}(H)$, then $\|\chi f\| = \|f\|$. In fact,

$$\begin{aligned}\|\chi f\| &= \sup_{\pi\in\widehat{H}} \|\pi(\chi f)\| = \sup_{\pi\in\widehat{H}} \|(\chi\pi) f\|\\ &= \sup_{\pi\in\widehat{H}} \|\pi(f)\| = \|f\|.\end{aligned}$$

We conclude from all this that, if $a = (b, \chi) \in N$, there is a unique isometry $v \mapsto av$ $(v \in \mathfrak{H})$ such that if $f \in \mathcal{L}(H)$, then we have

$$(af)(x) = (\chi(x) f(b^{-1}xb))/\Delta(b)$$

whence Lemma 1.

LEMMA 2. *For* $a = (b, \chi)$ *and* $\pi \in \widehat{H}$, *define*

$$(a\pi)(x) = \chi(x)\pi(b^{-1}xb)/\Delta(b).$$

We claim that N *acts on* $\widehat{H}$ *as a topological transformation group.*

In fact, we start by recalling (cf. [**D***], 3.9.9, p. 81) that if A is a separable C^*-algebra, N a topological group, and $\omega : N \to \mathrm{Aut}(A)$ is, under pointwise convergence, a continuous homomorphism then, setting for $\pi \in \widehat{A}$, $a \in N : (a\pi)(v) = \pi(\omega(a^{-1})v)$, N acts on $\widehat{A}$ as a topological transformation group. If $a = (b, \chi) \in N$, $a' = (b, \overline{\chi})$, $a \mapsto a'$ is a continuous automorphism of N. In this fashion it is enough to observe that $(a\pi)(v) = \pi((a')^{-1}v)$ $(a \in N, \pi \in \widehat{H}, v \in \mathfrak{H})$ whence the desired conclusion is implied by Lemma 1.

In the following we return to the notation as at the start. From what precedes we can infer

LEMMA 3. $\widetilde{G}$ *acts on* $\widehat{L}$ *as a topological transformation group via the action that is contragredient to conjugation.*

LEMMA 4. *We claim that* $\widehat{L}/\widetilde{G}$ *is countably separated.*

In fact, this is implied by the Main Lemma (cf. 1.3).

In the following we use the notation and results of 3.1. We note that, in particular, for any $\pi \in \widehat{L}$ and $a \in \widetilde{G}$ we have $K_{a\pi} = K_\pi$. By virtue of the last property, if $E \in \widehat{L}/\widetilde{G}$ we can write $K(E) \equiv K_\pi$ $(\pi \in E)$.

LEMMA 5. *Suppose that* $E \in \widehat{L}/\widetilde{G}$ *is fixed. We write* $K = K(E)$ *and* $F(E) = \{\rho; \rho \in \widehat{K}, \rho|L \in E\}$ *(or just* F*). We claim that in the topology induced by that of* $\widehat{K}$, F *is locally compact and Hausdorff.*

In fact,

a) We claim that E is locally closed in $\widehat{L}$. This is so, because

1) $L = (I)$ and thus L is postliminaire (cf. [**D***], 9.1. Théorème, p. 168, (i)⇒(iii)).

2) From 1) we conclude that $\widehat{L}$ is locally quasi-compact and almost Hausdorff (cf. [**D***], 4.5.7, p. 94, and [**Gl**], p. 125).

3) By Lemma 4, we infer that $\widehat{L}/\widetilde{G}$ is countably separated. By [**Gl**], Theorem 1 (3) ⇒ (2), p. 124, this implies that any $\widetilde{G}$-orbit E is open in its closure and thus E is locally closed for all $E \in \widehat{L}/\widetilde{G}$ as claimed above.

b) Next we show that F is locally closed in $\widehat{K}$. In fact, we recall (cf. 1.3, Lemma 1) that if G is separable, locally compact, which contains H as a closed, invariant, type I subgroup and $A \subset \widehat{H}$ is G-invariant and locally closed and $\widehat{G}_A = \{\rho; \rho \in \widehat{G}$ such that A carries $\rho|H\}$, then $\widehat{G}_A$ is locally closed in $\widehat{G}$. Hence if we replace G, H, and A respectively by K, L, and E, then $\widehat{K}_E$ is locally closed in $\widehat{K}$. In this manner it is enough to show that $F = \widehat{K}_E$. Assume that ρ is fixed in F. Since K leaves E pointwise fixed, there is $\pi \in E$ such that $\rho|L$ is quasi-equivalent to π. But since π extends to K, $\rho|L$ is unitarily equivalent to π.

c) Next we recall the following (cf. [**D***], p. 61). Let A be a separable C^*-algebra. Then there is a bijection between the family of all open subsets of $\widehat{A}$ and closed 2-sided ideals in A. In fact, if I is such, then the corresponding open set is $\widehat{I} = \widehat{A} - \widehat{A}_I$ ($\widehat{A}_I = \{\pi; \pi \in \widehat{A}, \ker(\pi) \supseteqq I\}$. We set $\mathcal{R} = C^*(K), \widehat{K} = \widehat{\mathcal{R}}$. By b), F is locally closed. Therefore there are $\widehat{K} \supset O_j$, $O_j =$ open in $\widehat{K}, O_1 \supset O_2$ such that $F = O_1 - O_2$. Hence, if I_j are ideals in $\mathcal{R}$ such that $O_j = \widehat{I}_j$ and $B = I_1/I_2$, then $F = \widehat{B}$.

d) We note next that B is postliminaire (cf. [**D***], 4.3.1, p. 87). In fact, to show this, it is enough to establish that if σ is arbitrary in $\widehat{B}$, then $\sigma(B)$ contains a nonzero compact operator. In fact, let us write also σ for (σ as above lifted to I_1). There is a representation ρ in F such that $\rho|I_1 = \sigma$. To attain our goal it is enough to show that there is $v \in \mathcal{R}$ such that $\rho(v) \neq 0$ and compact.

We set $\mathcal{L} = C^*(L)$ and write dy for the Haar measure on L. We note that $\mathcal{R}$ carries the structure of a left $\mathcal{L}$-module well-determined by the following.

If f is continuous and of a compact support on L, l is the left regular representation of L on $\mathcal{R}$, then for each $g \in \mathcal{R} : fg = \int_L f(y)l(y)g\,dy$. Note that if τ is some unitary representation of K, then for $f \in \mathcal{L}, g \in \mathcal{R}$, we have $\tau(f \cdot g) = \tau(f)\tau(g)$.

We recall now that $\pi = \rho|L$. Since L is of type I, there is $f \in \mathcal{L}$ ($= C^*(L)$) such that $\pi(f)$ is nonzero and compact. To complete our reasoning, it is enough to note that there is $g \in I$ such that $\sigma(fg)$ is nonzero.

Conclusion: We have shown that if $F = O_1 - O_2$, $O_k = \widehat{I}_k$ ($k = 1, 2$), $B = I_1/I_2$, then B is postliminaire, as claimed at the start of d). This permits us to infer that F, as a subspace of $\widehat{K}$, is locally quasi-compact and almost Hausdorff.

e) We write $\mathcal{F}$ for the annihilator of L in $\mathrm{X}(K)$, and set $N = \widetilde{G} \times \mathcal{F}$. We recall (cf. Lemma 3) that N acts on $\widehat{K}$ as a topological transformation group. Also, if $\rho \in F$ is fixed, then $F = N\rho$. Since F is locally quasi-compact and almost Hausdorff, by [**Gl**], Theorem 1, p. 123, the map $\iota : N/N_\rho \to F$ defined by $\iota(aN_\rho) = a\rho$ is a homeomorphism. To conclude our proof, it is enough to show that N_ρ is closed.

To see this, we note first that since $F = \widehat{B}$ and B is postliminaire, F is standard (cf. [**D***], 4.6.1, Proposition, p. 95) and thus so is N/N_ρ. This, however, implies that N_ρ is closed (cf. [**AM**], p. 16).

Summing up, $F = N/N_\rho$ is locally compact and Hausdorff, as claimed in Lemma 5.

Unexplained notation as above. We put

$$\mathcal{A} = \bigcup_{E \in \widehat{L}/\widetilde{G}} F(E) \quad \text{(disjoint union)}.$$

LEMMA 6. *There is an equivalence relation Σ on $\mathcal{A}$ such that $\rho_1, \rho_2 \in \mathcal{A}$ are equivalent if and only if there is $E \in \widehat{L}/\widetilde{G}$ such that $\rho_1, \rho_2 \in F(E)$ and $\rho_2 \in \overline{G\rho_1}$ (the closure being taken in $F(E)$). For any $E \in \widehat{L}/\widetilde{G}$, $F(E)/\Sigma$ is countably separated.*

In fact, let us show first that if ρ_1, ρ_2 are in $F(E)$ such that $\rho_2 \in \overline{G\rho_1}$ then $\overline{G\rho_1} = \overline{G\rho_2}$. Let ρ be fixed in $F(E)$. We have $[N, N] = [\widetilde{G}, \widetilde{G}] = L \subset K \subset N_\rho$. Hence N_ρ is invariant in N and N/N_ρ is abelian. Let us write again ι for the canonical homeomorphism $N/N_\rho \to F(E)$. We denote by A the closure of the image of G in N/N_ρ. All this being so, to obtain the desired conclusion it is enough to observe that $\iota(Am) = \overline{G\iota(m)}$ ($m \in N/N_\rho$).

Evidently, $F(E)/\Sigma$ is Hausdorff and hence (cf. [**AM**], p. 16) $F(E)$ is countably separated.

Below we follow closely the relevant parts of [**AM**], cf. in particular p. 83.

Below L will be assumed to be locally compact, separable, and of type I. Given a representation T of L, by abuse of language we shall quote as canonical measure one member of the measure class that T determines on $\widehat{L}$ (cf. [**D***], 8.6.5, p. 155). Since L is of type I, this is the same as the central measure of T.

Let T be a representation of L, and E a Borel set in $\widehat{L}$. We shall say that T is carried by E if $\widehat{L} - E$ is of measure zero with respect to the canonical measure.

In this case we shall write $T \subset E$.

LEMMA 7. *Assume that (Λ, μ) is a standard measure space and let $\lambda \mapsto T(\lambda)$ $(\lambda \in \Lambda)$ be a μ-measurable field of continuous unitary representations of L. Let us put $T = \int_\Lambda^\oplus T(\lambda) \cdot d\mu(\lambda)$. Then $T \subset E$ if and only if $T(\lambda) \subset E$ for almost all λ with respect to μ.*

In fact,

a) Let us assume that the ring of decomposition S is maximal in $(R(T))'$. Then we can assume the following situation. (X, τ) is a standard measure space, $\rho : X \to \mathrm{Irr}(L)$ is a Borel map, $T = \int_X^\oplus \rho(x) \cdot d\tau(x)$ and $S = L_\tau^\infty(X)$.

1) Let p be the canonical projection $\mathrm{Irr}(L) \to \widehat{L}$. Let us put $q(x) \equiv p(\rho(x))$; then q is a Borel map from X into $\widehat{L}$. We have $\rho(x) \subset E$ if and only if $q(x) \in E$. We put $B_0 = q^{-1}(\widehat{L} - E)$. Then, evidently, $\rho(x) \subset E$ for almost all x with respect to τ if and only if $\tau(B_0) = 0$.

2) Next we recall the following. Let us put $\nu = q_*(\tau)$. Then there is, up to a set of ν-measure zero, for each $\zeta \in \widehat{L}$ a measure τ_ζ on X, carried by $q^{-1}(\{\zeta\})$ such that for each Borel set $B \subset X$, $\zeta \mapsto \tau_\zeta(B)$ is ν-measurable and

$$\tau(B) = \int_{\widehat{L}} \tau_\zeta(B) \cdot d\nu(\zeta).$$

We set

$$U(\zeta) = \int_X^\oplus \rho(x) \cdot d\tau_\zeta(x).$$

Evidently, $U(\zeta)$ is quasi-equivalent to ζ.

Also (cf. [**Mackann II**], Theorem 2.11, p. 204), $\zeta \mapsto U(\zeta)$ is ν-measurable and we have

$$T = \int_X^\oplus \rho(x) \cdot d\tau(x) = \int_{\widehat{L}}^\oplus U(\zeta) \cdot d\nu(\zeta).$$

Since L is of type I, ν is the canonical measure of T. We conclude from this that T is carried by E if and only if $\nu(\widehat{L} - E) = 0$. This is the case if and only if $\tau(B_0) = 0$, which is equivalent to $\rho(x) \subset E$ almost everywhere with respect to τ.

b) This being so, we settle the general case in the following manner.

1) We assume that (Λ, μ) is standard, and

$$T = \int_\Lambda^\oplus T(\lambda) \cdot d\mu(\lambda).$$

Let P be the ring of decomposition; it identifies to $L^\infty_\mu(\Lambda)$. Let S be maximal abelian in the commuting ring of T, such that $P \subset S$. Assume that

$$T = \int_X^\oplus \rho(x) \cdot d\tau(x)$$

is a disintegration corresponding to S, which now identifies to $L^\infty_\tau(X)$. We can assume (cf. [**AM**], pp. 12 and 83) that there is a Borel map $h : X \to \Lambda$ such that $\mu = h_*(\tau)$ and $h^* : L^\infty_\mu(\Lambda) \to L^\infty_\tau(X)$ corresponds to the given inclusion.

2) Proceeding as in a) we can conclude that for each λ in Λ, there is a Borel measure τ_λ on X such that it is carried by $h^{-1}(\{\lambda\})$ and that for any Borel set $B \subset X$ we have

$$\tau(B) = \int_\Lambda \tau_\lambda(B) \cdot d\mu(\lambda).$$

Let us put $V(\lambda) = \int_X \rho(x) \cdot d\tau_\lambda(x)$. Then

$$T = \int_\Lambda^\oplus V(\lambda)\, d\mu(\lambda)$$

and $T(\lambda) \sim V(\lambda)$ for almost all λ with respect to μ.

3) α) Assume that T is carried by E; we are to show that $T(\lambda)$ is carried by E for almost all λ with respect to μ. We have by a): $\tau(B_0) = 0$ and thus $\tau_\lambda(B_0) = 0$ for almost all λ with respect to μ.

β) Let us assume now that $T(\lambda) \subset E$ for almost all $\lambda(\mu)$. We have to show that if

$$T = \int_\Lambda^\oplus T(\lambda) \cdot d\mu(\lambda)$$

then $T \subset E$. We have by a) for almost all $\lambda(\mu)$, $\tau_\lambda(B_0) = 0$ and therefore

$$\tau(B_0) = \int_\Lambda \tau_\lambda(B_0) \cdot d\mu(\lambda) = 0$$

whence again by a), $T \subset E$.

Summing up we have completed proving Lemma 7: If L is of type I, (Λ, μ) is standard and the map $\lambda \mapsto T(\lambda)$ is a μ-measurable field of representations, let us put

$$T = \int_\Lambda^\oplus T(\lambda) \cdot d\mu(\lambda).$$

Then T is carried by E if and only if $T(\lambda) \subset E$ for almost all λ with respect to μ.

Next below, because of later applications, we shall be more general than currently needed. We recall (cf. [**D***], 5.7.6, p. 114) that if A is a separable C^*-algebra, the representation T of A is called homogeneous if and only if for any subrepresentation T_1 of T we have $\ker(T_1) = \ker(T)$. We note that the kernel of any homogeneous representation is always a primitive ideal. For example, any factor representation of A is always homogeneous. We denote by $\hom(A)$ the totality of all homogeneous representations of A. In particular, if G is separable and locally compact, then we set

$$\hom(G) = \hom(C^*(G)).$$

We recall that if L is of type I, T some continuous unitary representation of L, $E =$ Borel $\subset \widehat{L}$, then T is carried by E if and only if the canonical measure T is carried by E. In this case we write $T \subset E$.

LEMMA 8. *Let G_1 be a closed subgroup of G such that $L \subseteqq G_1 \subseteqq G$. Let J be a fixed primitive ideal of G_1. Then there is an orbit $O \in \widehat{L}/\widetilde{G}$, well-determined by J, and such that if $T \in \hom(G_1)$ and $\ker(T) = J$, then $T|L$ is carried by O.*

1) In fact, we start by establishing the uniqueness. Assume that $O_k \in \widehat{L}/\widetilde{G}$ $(k = 1, 2)$ are such that $O_1 \neq O_2$ and, if $T \in \hom(G_1)$ is of kernel J, then both carry $T|L$. Let μ be the canonical measure of $T|L$. If O_1 carries $T|L$, then $\mu(\widehat{L} - O_1) = 0$ and thus $\mu(O_2) = 0$. By symmetry, we have also $\mu(O_1) = 0$.

2) We turn now to the question of existence. If T is a factor representation of G_1, then the canonical measure of $T|L$ $(= \mu$, say; we can assume $\mu(\widehat{L}) = 1)$ is ergodic with respect to the action of G_1 on $\widehat{L}$. We recall (cf. Lemma 4 above) that $\widehat{L}/\widetilde{G}$ is countably separated. We observe that there is a $\widetilde{G}$-orbit, well-determined by the condition that it carry μ. In fact, our hypothesis implies that there is a sequence (O_k) of $\widetilde{G}$-invariant Borel sets, such that if O is any $\widetilde{G}$-orbit, then

$$O = \bigcap_k{}' O_k$$

(prime indicates restriction to all those O_k's for which $O \subseteqq O_k$). We define a sequence (A_j) by the following stipulation: $A_j = O_j$ if $\mu(O_j) = 1$ and $A_j = CO_j$ otherwise. Let us set $A = \bigcap_j A_j$. We have then $\mu(A) = 1$, and A is $\widetilde{G}$-invariant.

In this manner it is enough to show that A is a $\widetilde{G}$-orbit. In fact, assume that $p, q \in A$ are such that $\widetilde{G}p \neq \widetilde{G}q$. Then there is a j such that $p \in O_j$, $q \notin O_j$, say, and thus p and q cannot be both in A. We conclude in this fashion that there is $O \in \widehat{L}/\widetilde{G}$, well-determined by the condition that it carries $T|L$.

Assume that $T_k \in \mathrm{Fac}(G_1)$ are such that $\ker(T_1) = J = \ker(T_2)$. Assume that $T_k|L$ is carried by $O_k \in \widehat{L}/\widetilde{G}$ $(k = 1, 2)$. We claim that $O_1 = O_2$. In fact, let μ_k be the canonical measure corresponding to $T_k|L$ on $\widehat{L}$ $(k = 1, 2)$. Given some measure μ on $\widehat{L}$ we denote by $h(\mu)$ its closed hull. This being so we can claim first that $h(\mu_1) = h(\mu_2)$. Since $\ker(T_1) = \ker(T_2)$, then T_1 and T_2 are weakly equivalent. But then so are $T_1|L$ and $T_2|L$ implying that $h(\mu_1) = h(\mu_2)$.

Let us assume that $O_1 \neq O_2$; we shall show that this is absurd. In fact,

1) We have $0 \leq \mu_2(C\overline{O}_2) \leq \mu_2(CO_2) = 0$ and hence $h(\mu_2) \subseteqq \overline{O}_2$.

2) By $h(\mu_1) = h(\mu_2)$, $h(\mu_1) \subseteqq \overline{O}_2$.

3) We observe that $\widehat{L}/\widetilde{G}$ is a T_0-space. In fact, by Lemma 4 above, $\widehat{L}/\widetilde{G}$ is countably separated and thus the desired conclusion is implied by (3) $\Rightarrow$ (2) in [**Gl**], Theorem 1, p. 124.

4) In this manner, we can assume that there is an open set $O \subset \widehat{L}$ such that $0_1 \subset O, O \cap \overline{O}_2 = \varnothing$ (say). We conclude from this that $0 = \mu_1(CO_1) \geqq \mu_1(CO) \geqq \mu_1(\overline{O}_2) \geqq \mu_1(h(\mu_1))$ implying $\mu_1(h(\mu_1)) = 0$ which is absurd. Summing up from what we have seen so far, we can conclude that if $J \in \mathrm{Prim}(G_1)$, then there is $O \in \widehat{L}/\widetilde{G}$ well-determined by: If T is a factor representation of G_1 such that $\ker(T) = J$, then $T|L$ is carried by O. To settle the general case we have to show that if $T \in \hom(G_1)$ is such that $\ker(T) = J$, then $T|L$ is carried by O (the latter

as before). We recall (cf. [**E**], Lemma 1.9, p. 91) that if

$$T = \int_{\Lambda}^{\oplus} T(\lambda) \cdot d\mu(\lambda)$$

is a decomposition into factor representations then $\ker(T(\lambda)) \equiv J$ a.e. (μ). Hence we have first

$$T|L = \int_{\Lambda}^{\oplus} (T(\lambda)|L) \cdot d\mu(\lambda)$$

and next, by Lemma 7, $T|L \subset O$, which completes the proof of Lemma 8.

Below, given $J \in \mathrm{Prim}(G)$, we shall write $E(J)$ for O (as in Lemma 8). If J is specified by the context, we may drop it altogether from our notation.

We recall that

$$\mathcal{A} = \bigcup_{E \in \widehat{L}/\widetilde{G}} F(E).$$

If $\rho \in \mathcal{A}$ is such that $\rho|L = \pi \in \widehat{L}$, then we write $T(\rho) = \mathrm{ind}_{K_\pi \uparrow G}\, \rho$ and we put $J(\rho) = \ker(T(\rho))$. We note that since $T(\rho)$ is a factor representation, $J(\rho)$ is a primitive ideal of G (cf. [**Dal**], Corollaire 3, p. 100). In this manner we have defined a map $J : \mathcal{A} \to \mathrm{Prim}(G)$.

LEMMA 9. *Assume that ρ_1, ρ_2 in $\mathcal{A}$ are such that $J(\rho_1) = J(\rho_2)$. We claim that ρ_1, ρ_2 lie on the same Σ-orbit.*

In fact, let ρ in $\mathcal{A}$ be such that $J(\rho) = J$. Then ρ is in $F(J)$. This is so, since Lemma 8 implies that $T(\rho)$ is carried by $E(J)$. In this manner, if $J(\rho_1) = J = J(\rho_2)$, then we have $\rho_j \in F(J)$ $(j = 1, 2)$. For more about what follows, see the proof of Lemma 5. We write $K = K(J)$ and $F = F(J)$. F is locally closed in $\widehat{K}$. Hence there are closed 2-sided ideals I_1, I_2 in $C^*(K)$ such that $I_1 \supset I_2$ and $F = \widehat{I_1} - \widehat{I_2}$. Therefore, if $B = I_1/I_2$, then $F = \widehat{B}$ and B is of type I. We have

$$T(\rho_k)|K = \int_{G/K}^{\oplus} \sigma\rho_k \cdot d\sigma.$$

We write T_k' for the representation corresponding to $T(\rho_k)|K$ of B. By virtue of our assumption $\ker(T_1') = \ker(T_2')$, which implies $\overline{G\rho_1} = \overline{G\rho_2}$ in $\widehat{K}$. Hence ρ_1 and ρ_2 lie on the same Σ-orbit.

LEMMA 10. *Assume that ρ_1, ρ_2 in $\mathcal{A}$ lie on the same Σ-orbit. We claim that $J(\rho_1) = J(\rho_2)$.*

In fact, our assumption implies that $\ker(T(\rho_1)|K) = \ker(T(\rho_2)|K)$. We recall (cf. 2.7) that if U_1, U_2 are representations of K such that $\ker(U_1) = \ker(U_2)$, then

$$\ker\Big(\mathrm{ind}_{K\uparrow G}\, U_1\Big) = \ker\Big(\mathrm{ind}_{K\uparrow G}\, U_2\Big).$$

Since, evidently, $T(\rho_k)$ is quasi-equivalent to $\mathrm{ind}_{K\uparrow G}(T(\rho_k)|K)$ $(k = 1, 2)$ we get finally $J(\rho_1) = J(\rho_2)$.

Summing up, by what we have seen in the last two lemmas, the “orbits” of the map $J : \mathcal{A} \to \mathrm{Prim}(G)$ coincide with those of the equivalence relation Σ.

3.3. Proof that J is surjective

Below our objective will be to show that the map $J : \mathcal{A} \to \mathrm{Prim}(G)$ is surjective. In doing so, we shall follow closely 2.4. We start by recalling the following proposition (cf. Lemma, loc. cit.). Assume that G is separable and locally compact and that (X, μ) is an ergodic Borel G-space. Assume also that $V : X \to \mathrm{Fac}(G)$ is μ-measurable such that if $Gx = Gx'$, then $V(x)$ is unitarily equivalent to $V(x')$. Let us form

$$V = \int_X^{\oplus} V(x) \cdot d\mu(x).$$

Then V is a factor representation.

LEMMA 1. *Assume that J is some element of* $\mathrm{Prim}(G)$. *We claim that there is a ρ in $\mathcal{A}$ such that $J(\rho) \subseteqq J$.*

Below we write just F in place of $F(J)$.

a) Let T be a factor representation of G such that $\ker(T) = J$. We write

$$T' = \mathrm{ind}_{K \uparrow G}(T|K)$$

and claim that T' is a factor representation of G such that for some $\rho \in \mathcal{A}$: $\ker(T') \supseteqq J(\rho)$. To see this

1) Recall (cf. [**D***], 8.4.2, p. 149) that there is a standard measure μ on $\widehat{K}$ and a μ-measurable field $(T(\zeta))$ of factor representations such that $T(\zeta)$ is of the class $\zeta \in \widehat{K}$,

$$T|K = \int_{\widehat{K}}^{\oplus} T(\zeta) \cdot d\mu(\zeta),$$

and that $(R(T|K))^{\mathrm{c}}$ corresponds to the totality of diagonalizable operators. We write E for $E(J)$ and recall (cf. 3.2, Lemma 8) that, by $\ker(T) = J$, $T|L$ is carried by E. Let A be the set

$$\{\zeta; \zeta \in \widehat{K}, T(\zeta)|L \subseteqq E\}.$$

We note that $\mu(CA) = 0$. In fact, since

$$T|L = \int_{\widehat{K}}^{\oplus} (T(\zeta)|L) \cdot d\mu(\zeta)$$

and $T|L \subseteqq E$, the desired conclusion follows from Lemma 7 in 3.2.

2) Let us denote by F' the image of F in $\widehat{K}$, under the canonical isomorphism of $\widehat{K}$ with its image $\widehat{K}_I$. Observe that evidently A is contained in F'. By taking inverse images, we obtain the central decomposition

$$T|K = \int_F^{\oplus} \sigma(\lambda) \cdot d\nu(\lambda)$$

where $\sigma(\lambda)$ is a multiple of λ ($\lambda \in F$), ν is quasi-invariant under the action of G on F and, since T is a factor representation, it is also ergodic. Let us set for $\lambda \in F$

$$V(\lambda) = \mathrm{ind}_{K \uparrow G}\, \sigma(\lambda).$$

We have (cf. [**Mackann I**], Theorem 10, p. 123)

$$T' = \mathrm{ind}_{K \uparrow G}(T|K) = \int_F^{\oplus} V(\lambda) \cdot d\nu(\lambda).$$

We note that $V(\lambda)$ is a factor representation. If $G\lambda = G\lambda'$, then $\sigma(\lambda)$ is unitarily equivalent to $\sigma(\lambda')$, and hence so is $V(\lambda)$ to $V(\lambda')$. In this manner we conclude that T' is a factor representation of G. We have also that, for almost all λ with respect to ν, $\ker(T') \supseteqq \ker(V(\lambda))$ (cf. [**E**], Lemma 1.9, p. 91). This being so, it is sufficient to observe that $\ker(V(\lambda))$ is of the form $J(\rho)$ $(\rho \in F)$.

b) To complete our proof of Lemma 1, it is enough to note the following result, the proof of which is obtained by a trivial modification of the reasoning of 2.4.

Suppose that J is given in $\mathrm{Prim}(G)$ and T is a factor representation of G such that $\ker(T) = J$. Let us put $T' = \mathrm{ind}_{K\uparrow G}(T|K)$. Then $\ker(T) \supseteqq \ker(T')$.

For a proof of the following Lemma 2, we refer to 2.9.

LEMMA 2. *Let Γ be a central extension, by the circle group, of a closed subgroup of an n-dimensional real vector space, and J a primitive ideal, of class one, of Γ. Then if U is the centralizer of $(\Gamma_0)^{\mathrm{c}}$, there is a character χ, of class one, of U^{c}, such that*

$$J = \ker\left(\mathrm{ind}_{U^{\mathrm{c}}\uparrow\Gamma}\,\chi\right).$$

A similar statement holds true for the class minus one case.

LEMMA 3. *Assume that J is a given element of* $\mathrm{Prim}(G)$. *We consider the following two conditions*:

(1) *J is of the form $J(\rho)$, where ρ is some element of $\mathcal{A}$.*

(2) *There is a factor representation V of G such that* $\ker(V) = J$ *and $V|L$ is transitive on $\widehat{L}$.*

We claim that (1) *and* (2) *are equivalent.*

We show first that (1) implies (2). To this end let us assume that J is of the form $J(\rho)$, with ρ in $\mathcal{A}$. We define

$$V = \mathrm{ind}_{K\uparrow G}\,\rho.$$

Then, by definition, $J = \ker(V)$ and, evidently, $V|L$ is transitive.

Let us show next that (2) implies (1). Assume that $V|L$ is of the form $G\pi$. Let π^e be a projective extension of π to G_π, and G_π^{e} the corresponding central extension of G_π. We form the Mackey group $M_\pi = G_\pi^{\mathrm{e}}/L$; let ϕ be the canonical morphism of G_π^{e} onto M_π. Then M_π is as Γ in Lemma 2. Let $\mathcal{T}$ and T be the canonical circle groups in $(M_\pi)^{\mathrm{c}}$ and $(G_\pi^{\mathrm{e}})^{\mathrm{c}}$ respectively.

Below we shall use the following notational convention. If U is some representation of G_π^{e} such that $U|T$ is constant, then $U^\sim$ will stand for the corresponding representation of $G_\pi^{\mathrm{e}}/T \sim G_\pi$. Since $V|L$ is transitive, there is $\tau \in \mathrm{Fac}(M_\pi)$ of class minus one, such that

$$V = \mathrm{ind}_{G_\pi\uparrow G}(\pi^e \otimes (\tau\circ\phi))^\sim.$$

By Lemma 2 above, there is a χ in $\mathrm{X}(U^e)$ such that if $V(\chi) = \mathrm{ind}_{U^{\mathrm{c}}\uparrow M_\pi}\,\chi$, then $\ker(\tau) = \ker(V(\chi))$.

Let us put

$$V' = \mathrm{ind}_{G_\pi\uparrow G}(\pi^e \otimes (V(\chi)\circ\phi))^\sim.$$

By the proof of Lemma 8 in 3.1, there is ρ in $\mathcal{A}\cap\widehat{K}_\pi$ such that $V' = \mathrm{ind}_{K_\pi\uparrow G}\,\rho$.

In this manner it is enough to point out that $\ker(V) = \ker(V')$. But this follows at once from 2.7.

LEMMA 4. *Assume that* $J \in \mathrm{Prim}(G)$ *is given. We suppose that there is a* $V \in \mathrm{Fac}(G)$ *such that* $\ker(V) \supseteqq J$ *and that* $V|L$ *is transitive, contained in* $E(J)$. *We claim that, for some* ρ *in* $\mathcal{A}$, J *is of the form* $J(\rho)$.

In fact, let us write $J_1 = \ker(V)$ and note that by $V \in \mathrm{Fac}(G)$, J_1 lies in $\mathrm{Prim}(G)$. Next we claim that $E(J) = E(J_1)$. In fact,

1) We have by assumption: $V|L \subseteqq E(J)$.

2) By Lemma 8 in 3.2, $V|L \subseteqq E(J_1)$ and thus $E(J) = E(J_1)$.

Next we note that, by Lemma 3, there is an element ρ_1 in $F(J_1)$ such that $J_1 = J(\rho_1)$. Lemma 1 supplies a ρ_2 in $\mathcal{A}$ such that $J \supseteqq J(\rho_2)$. In this fashion we have $J(\rho_1) \supseteqq J \supseteqq J(\rho_2)$. Hence it is enough to show that $J(\rho_1) = J(\rho_2)$. To see this, let us put $T(\rho_j) = \mathrm{ind}_{K\uparrow G}\, \rho_j$ $(j = 1, 2)$. Our assumption implies that $\ker(T(\rho_1)) \supseteqq \ker(T(\rho_2))$ and thus also $\ker(T(\rho_1)|K) \supseteqq \ker(T(\rho_2)|K)$ and $\overline{G\rho_1} \supseteqq \overline{G\rho_2}$. Hence ρ_1 and ρ_2 lie on the same Σ-orbit in which manner, by Lemma 10 in 3.2., we can conclude that $J(\rho_1) = J(\rho_2)$.

By virtue of what we have just seen, the surjectivity of the map $J : \mathcal{A} \to \mathrm{Prim}(G)$ will follow from

LEMMA 5. *Suppose that* J *is a given primitive ideal of* G. *We claim that there is a factor representation* V *of* G *such that* $\ker(V) \supseteqq J$ *and* $V|L$ *is transitive, carried by* $E(J)$.

In fact, given $\pi \in E(J)$, we put $G(J) = G_\pi$.

a) We recall first that by Lemma 8 in 3.2, if T is irreducible, such that $\ker(T) = J$, then $T|L \subseteqq E(J)$. Also, if $j : \widetilde{G}/\widetilde{G}_\pi \to E(J)$ is defined by $j(a\widetilde{G}_\pi) = a\pi$ $(a \in \widetilde{G})$ then j is a $\widetilde{G}$-equivariant homeomorphism. In fact, since L is of type I, $\widehat{L}$ is locally quasi-compact and almost Hausdorff. This is acted upon by $\widetilde{G}$ in a countably separated manner (cf. Lemma 4 in 3.2); thus it is enough to invoke (3) $\Rightarrow$ (6) in [**Gl**], Theorem 1, p. 124. In this fashion $T|L$ is carried by a relative G-orbit closure in $E(J)$. All this being so, we can assume to be in the following situation. There is given a connected abelian Lie group $\mathfrak{t}$ and a morphism φ of G into $\mathfrak{t}$ such that $\ker(\varphi) = G(J)$. Putting $\mathfrak{t}' = \varphi(G)$ we have $\overline{\mathfrak{t}'} = \mathfrak{t}$. We are given also a G-equivariant Borel injection $\pi : \mathfrak{t} \to \mathrm{Irr}(L)$ and a Borel measure μ on $\mathfrak{t}$ such that

$$(+) \qquad T|L = \int_{\mathfrak{t}}^{\oplus} \pi'(\zeta) \cdot d\mu(\zeta)$$

where $\pi'(\zeta) \approx \pi(\zeta)$ (cf. [**D***], 4.6.2, Proposition, p. 95).

b) Below we shall use the following elementary fact. Let $\mathfrak{t}$ be a connected abelian Lie group and $\mathfrak{t}'$ a connected dense subgroup. Then there is a connected closed subgroup $\mathfrak{t}_1$ of $\mathfrak{t}$ such that $\mathfrak{t} = \mathfrak{t}_1 \cdot \mathfrak{t}'$ and, if $\mathfrak{t}_1' = \mathfrak{t}_1 \cap \mathfrak{t}'$, then $\mathfrak{t}_1'$ is countable.

c) We write $\mathcal{B}$ for the quotient $\mathfrak{t}/\mathfrak{t}_1$ and denote by ω the corresponding canonical morphism. Then $\omega' = \omega \circ \varphi$ is a continuous morphism of G onto $\mathcal{B}$. If S is its kernel, then $S/G(J)$ is discrete. We claim that there is an irreducible representation T_1 of S such that $T = \mathrm{ind}_{\rho\uparrow G}\, T_1$. In fact, let P be a spectral measure, corresponding to (+), with values in $(R(T|L))^{\mathrm{c}}$ on $\mathfrak{t}$. If we write Q for the direct image through ω' of P in $\mathcal{B}$, the pair (Q, T) provides a transitive system of imprimitivity of G/S. Hence, by [**Mackact**], Theorem 6.6, p. 281, we obtain an irreducible representation T_1 of S such that $T = \mathrm{ind}_{\rho\uparrow G}\, T_1$.

d) We note that $L \subseteqq G(J) \subseteqq S \subseteqq G$. Let $\mathcal{F}$ be the primitive ideal corresponding to T_1, in $\mathrm{Prim}(S)$. By Lemma 8 in 3.2, there is an element $E(\mathcal{F}) \in \widehat{L}/\widetilde{G}$, which carries $T_1|L$; it is uniquely determined by this property. Next we claim that $E(\mathcal{F}) = E(J)$. In fact, by $T = \operatorname{ind}_{\rho\uparrow G} T_1$ we have

$$T|L = \int_{G/S}^{\oplus} \tau(T_1|L) \cdot d\tau$$

from which the desired conclusion is implied by Lemma 7 in 3.2.

e) Let us put $H = S_0$. We observe again that if V_1 and V_2 are representations of S such that $\ker(V_1) = \ker(V_2)$, then we have also $\ker(V_1|H) = \ker(V_2|H)$. In this fashion $\mathcal{F}$ defines unambiguously a closed 2-sided ideal I of $C^*(H)$ for which we shall write $\mathcal{F}|H$.

Next we claim that there is a primitive ideal σ in $C^*(H)$ such that $I \subseteqq \sigma$ and $E(\sigma) = E(J)$. To show this, we shall employ the following result: there is a Borel measure ν on $\mathrm{Prim}(H)$ and a ν-measurable map $\rho : \mathrm{Prim}(H) \to \hom(H)$ such that $\ker(\rho(\lambda)) \equiv \lambda$ and

$$T_1|H = \int_{\mathrm{Prim}(H)}^{\oplus} \rho(\lambda) \cdot d\nu(\lambda)$$

(cf. [**E**], p. 97, top). We have also the following result (cf. [**E**], Lemma 3.4, p. 99). If $h(\nu)$ is the closed hull of ν and $\ker(h(\nu)) = \bigcap_{\lambda \in h(\nu)} \ker(\lambda)$, then

$$I = \ker(T_1|H) = \ker(h(\nu)).$$

Since $E(\mathcal{F}) = E(J)$, the desired conclusion is implied again by Lemma 7 in 2.3. Note that $\ker(\rho(\sigma)) \equiv \sigma$; whence $\rho(\sigma)|L \subseteqq E(\sigma)$. But we have also

$$T_1|H = \int_{\mathrm{Prim}(H)}^{\oplus} \rho(\lambda) \cdot d\nu(\lambda)$$

$$\Rightarrow T_1|L = \int_{\mathrm{Prim}(H)}^{\oplus} \rho(\lambda)|L \cdot d\nu(\lambda).$$

By $T_1|L \subseteqq E(J)$, we have $\rho(\sigma)|L \subseteqq E(J)$ implying $E(\sigma) = E(J)$.

f) Let U be an irreducible representation of H such that $\ker(U) = \sigma$. We claim that $U|L$ is a multiple of some π in $E(J)$. In fact, here it is sufficient to recall that, by $H = S_0$, H acts trivially on $E(J)$.

g) Before proceeding, we recall the following facts from 2.6. (We shall use notation as in loc. cit.)

1) Suppose that F is an open, invariant subgroup in G. Let J be a proper closed 2-sided ideal in $C^*(G)$. Putting $I = J|F$, assume that η is some element of EP_I; then there is ω in EP_J such that $\omega|F = \eta$ (cf. Lemma 2, loc. cit.).

2) Suppose that F is closed and invariant in G. We assume that $\omega \in P(G)$ is such that $\omega|F \in EP(F)$. We set $\eta = \omega|F$. Then there is a countable subset A of G such that, in the sense of unitary equivalence, we have

$$L^{\omega}|F = \bigoplus_{a \in A} a(L^{\eta})$$

(cf. Lemma 3 loc. cit.).

h) We apply g)1) in the following situation. Replace G by S, F by H. Let φ be a positive definite function on H and $\omega \in EP_{\mathcal{F}}$ be such that $\omega|H \equiv \varphi$. Let us put $L^{\omega} = Z$ and form $V' = \operatorname{ind}_{S\uparrow G} Z$. Assume that $U \in \mathrm{Irr}(H)$. Let $\pi \in E(J)$ be again such that $U|L$ is quasi-equivalent to π. This being so we claim

1) $\ker(V') \supseteqq J$;

2) $V'|L$ is carried by $G\pi$.

In fact,

1) By $\omega \in P_{\mathcal{F}} : \ker(Z) \supseteqq \mathcal{F} = \ker(T_1)$. By $T = \operatorname{ind}_{S\uparrow G} T_1 : \ker(V') \supseteqq J$.

2) We recall that $V' = \operatorname{ind}_{S\uparrow G} Z$, $Z = L^\omega$. We have to show that $V'|L$ is carried by $G\pi$. To this end, we note that by construction

$$V'|L = \int_{G/S}^{\oplus} \tau(Z|L) \cdot d\tau.$$

In this fashion we can conclude that $V'|L$ is carried by $G\pi$.

k) 1) We can write

$$V' = \int_{\operatorname{Prim}(G)}^{\oplus} \sigma(\lambda) \cdot d\mu(\lambda)$$

where $\sigma(\lambda)$ is homogeneous such that $\ker(\sigma(\lambda)) \equiv \lambda$. We infer from this that there is $W \in \hom(G)$ such that $\ker(V') \subseteqq \ker(W)$. Evidently $W|L \subseteqq E(J)$.

2) Let us form the central decomposition

$$W = \int_{\Lambda}^{\oplus} T(\lambda) \cdot d\alpha(\lambda).$$

By [**E**], 1.9, p. 91, $\ker(W) = \ker(T(\lambda))$ almost everywhere with respect to α. We conclude from this that, for an appropriate choice of $\lambda_0 \in \Lambda$, if $V = T(\lambda_0)$, then $\ker(V) = \ker(W) \supseteqq J$ and, evidently, $V|L$ is carried by $G\pi$, completing the proof of Lemma 5.

We summarize our previous discussion in the following

PROPOSITION. *Given $\rho \in \mathcal{A}$ such that $\rho|L = \pi \in \widehat{L}$, let us write*

$$J(\rho) = \ker\Big(\operatorname{ind}_{K_\pi\uparrow G} \rho\Big) \in \operatorname{Prim}(G).$$

(i) *The map $J : \mathcal{A} \to \operatorname{Prim}(G)$ is surjective.*

(ii) *We have $J(\rho_1) = J(\rho_2)$ if and only if ρ_1 and ρ_2 are on the same Σ-orbit of $\mathcal{A}$.*

(iii) *For a fixed $J \in \operatorname{Prim}(G)$ let us set $\mathfrak{t}(J) = \{\rho; \rho \in \mathcal{A}, J(\rho) = J\}$. Then $\mathfrak{t}(J)$ as a subspace of the dual of $K(J)$ is locally compact and Hausdorff. Given a factor representation V of G with a kernel equal to J there is an (up to equivalence) unique Radon measure μ on $\mathfrak{t}(J)$ quasi-invariant and ergodic with respect to the action of G, such that $V|K(J)$ is quasi-equivalent to*

$$\int_{\mathfrak{t}(J)}^{\oplus} \rho \cdot d\mu(\rho).$$

3.4. Technical tools

The objective of this section will be the discussion of some technical tools needed for the proof in Section 3.5 of the existence of normal representations with given kernels. To accomplish this, we start by suggesting some new notational conventions. G will continue to stand for a connected and simply connected Lie group; we put $L = [G, G]$. Given some Lie group H, we often write $\mathfrak{H}$ to denote its group C^*-algebra. $\mathcal{L}(H)$ will stand for the space of all continuous functions with compact support on H. This being so, given l in L, we denote by $R(l)$ the value of the right-regular representation of L on $\mathcal{L}$ $(= C^*(L))$.

We denote by $\mathcal{B}$ the collection of all continuous maps φ from G into $\mathcal{L}$, satisfying the following conditions:

1) For $l \in L$ and $x \in G$ we have $\varphi(lx) = R(l)\varphi(x)$;

2) $x \mapsto \|\varphi(x)\|$ is of compact support on $H = G/L$. Given $\varphi \in \mathcal{B}$, we write $\mathrm{supp}(\varphi)$ to denote the support of $xL \mapsto \|\varphi(x)\|$ on H.

As an example, let $g \in \mathcal{L}(G)$ be given and let us define, for given $x \in G$, an element $\varphi(x)$ of $\mathcal{L}$ by $(l \mapsto g(lx)) \in \mathcal{L}(L) \subset \mathcal{L}$. This being so, we note that φ is in $\mathcal{B}$. In fact, if $l_0 \in L$ is fixed, we have $\varphi(l_0x) = (l \mapsto g(ll_0x)) = R(l_0)\varphi(x)$. We write $\mathcal{B}_0$ for the collection of all elements, as just defined, of $\mathcal{B}$. Given elements (f_n), $f \in \mathcal{B}$, we say that (f_n) tends to f in $\mathcal{B}$ iff there is a compact set $K \subset H$ such that $\mathrm{supp}(f_n) \subset K$ and $f_n \to f$ uniformly.

LEMMA 1. *The natural map $\mathcal{L}(G) \to C^*(G)$ extends to a continuous injection $\mathcal{B} \to C^*(G)$.*

In fact, we write dx for the element of a right-invariant Haar measure on G, and dl, dh Haar measures on L and H resp. Given $f \in \mathcal{B}$, we set $[f] = \int_H \|f(x)\| \cdot dh$. We have, evidently, $[f] \leqq h(\mathrm{supp}(f)) \cdot \|f\|_\infty$.

a) Let $f \in \mathcal{L}(G)$ and $\varphi \in \mathcal{B}_0$ correspond to each other as above. If $\|\ \|_1$ is the norm on $C^*(G)$ we have $\|f\|_1 \leqq [\varphi]$. In fact, given $U \in \mathrm{Rep}(G)$ we have

$$\begin{aligned} U(f) &= \int_G f(x)U(x)\,dx = \int_H \left(\int_L f(lx)U(l)\,dl\right) U(x)\cdot dh \\ &= \int_H U(\varphi(x))\cdot U(x)\cdot dh \end{aligned}$$

and $\|U(x)\| \leqq \int_H \|\varphi(x)\| \cdot dh = [\varphi]$.

By letting U go over $\mathrm{Rep}(G)$ we conclude that $\|f\|_1 \leqq [\varphi(x)]$ $(f \in \mathcal{L}(G); f \leftrightarrow \varphi \in \mathcal{B}_0)$.

b) We note next that $\mathcal{B}_0$ is dense in $\mathcal{B}$. In fact, let H' be a closed submanifold of G such that $G = L{\cdot}H'$ setwise. If $f \in \mathcal{B}$, it is well defined by $f|H'$. Consider $C^*(L)\otimes \mathcal{L}(H')$ in the inductive topology; then $\varphi \mapsto f|H'$ gives rise to an isomorphism $\mathcal{B} \to C^*(L) \otimes \mathcal{L}(H')$. Let $g \in \mathcal{L}(L), m \in \mathcal{L}(H')$ be given, and define $f \in \mathcal{L}(G)$ by $f(lh) = g(l)m(h)$. We have then $\varphi(h) = (l \mapsto g(l)m(h))$ and thus it is enough to observe that $\mathcal{L}(L)$ is dense in $C^*(L)$.

c) We conclude by b) and $\|f\|_1 \leqq [\varphi]$ (cf. a) end) that there is a continuous map $\mathcal{B} \to C^*(G)$ extending $\mathcal{L}(G) = \mathcal{B}_0 \to C^*(G)$. This being so, if $f \in \mathcal{L}(G)$, we shall write also f in place of φ as above. For later use we note that if $f \in \mathcal{B}$, then $\|f\|_1 \leqq [f]$.

d) Given f in $\mathcal{B}$, we write f' for its image in $C^*(G)$. We claim that the map $f \mapsto f'$ is injective. In fact, we have to show that $f' = 0$ implies $f = 0$.

1) We note that if $f \in \mathcal{B}$ and $U \in \mathrm{Rep}(G)$ then $U(f) = \int_H U(f(x)) \cdot U(x) \cdot dh$. To see this we remark that this is valid if $f \in \mathcal{B}_0$, and thus, by b), it can be extended to the whole of $\mathcal{B}$.

2) This having been said, it is enough to show that if f in $\mathcal{B}$ is such that $\int_H U(f(x)) \cdot U(x) \cdot dh = 0$ for all $U \in \mathrm{Rep}(G)$, then $f(x) \equiv 0$. In fact, by replacing U through χU $(\chi \in \mathrm{X}(G))$, we can conclude that $U(f(x)) \equiv 0$ for all $U \in \mathrm{Rep}(G)$. Hence it suffices to note that if $v \in C^*(L)$ is such that $U(v) = 0$ for all $U \in \mathrm{Rep}(G)$, then $v = 0$. In fact, let $\pi \in \mathrm{Irr}(L)$ be such that $U = \mathrm{ind}_{L\uparrow G}\, \pi$ and thus $0 = U(v) = \int_{G/L} (h\pi)(v) \cdot dh$, whence $v = 0$.

Summing up, we have completed proving that the natural map $\mathcal{B}_0 \to \mathfrak{G}$ extends to a continuous injection $\mathcal{B} \to \mathfrak{G}$.

In the following we shall denote by the same letter the corresponding elements in $\mathcal{B}$, $C^*(G)$ and (if it applies) in $\mathcal{L}(G)$. As we have seen we have for $l \in \mathcal{B}$, and $U \in \operatorname{Rep}(G)$

$$U(f) = \int_H U(f(x)) \cdot U(x)\, dh.$$

Also, $\|f\|_1 \leqq [f]$ ($\| \ \|_1 =$ norm in $C^*(G)$ for which below we shall write just $\| \ \|$).

Next we shall show that $\mathcal{B} \cdot \mathcal{B} \subseteqq \mathcal{B}$ and $\mathcal{B}^* = \mathcal{B}$. Before doing that, however, we wish to make the following remark. We recall that a unitary representation of G is called normal if it generates a semifinite factor such that there is an element a of $\mathfrak{G}$ for which the trace of a^*a is positive and finite. We shall show later that in this case a can be chosen in $\mathcal{B}$. An improvement of this, due to J. Y. Charbonnel, claims that a can even be chosen to be some smooth function with compact support.

Let $R(L)$ stand for the right (left) regular representation of L on $\mathcal{L}$ $(= C^*(L))$. Assume that the Haar measure dx on G is such that $d(sx) = \Delta(s) \cdot dx$. Given $z \in G$, we denote by $A(z)$ the automorphism of $\mathcal{L}$ for which on $f \in \mathcal{L}(L)$ we have $(A(z)f)(x) \equiv f(z^{-1}xz)/\Delta(z)$. For $f \in C^*(L)$ we write f^s for the action of the canonical conjugation on f.

Below we shall use the following relations, the verification of which we leave to the reader. We have for $z \in G, l \in L$; $\alpha, \beta \in C^*(G)$:

1) $A(l) = L(l)R(l)$,
2) $A(z)L(l) = L(zlz^{-1})A(z)$,
3) $R(l)\alpha \cdot L(l)\beta = \alpha \cdot \beta$,
4) $R(l)(\alpha \cdot \beta) = \alpha \cdot R(l)\beta$,
5) $(R(l)\alpha)^s = L(l)\alpha^s$.

Let $f, g \in \mathcal{B}$, $x \in G$ be fixed and consider the following expression: $F(y) \equiv f(xy^{-1})A(xy^{-1})g(y)$. We claim that for $l \in L$ fixed we have the following relation: $F(ly) \equiv F(y)$.

In fact, we see at once that

$$F(ly) = f(xy^{-1}l^{-1})A(x \cdot y^{-1})A(l^{-1})A(l^{-1}),$$

which, by $A(l^{-1})R(l) = L(l^{-1})$ is equal to

$$\begin{aligned} F(ly) &= R(xy^{-1}l^{-1}(xy)^{-1}) = f(xy^{-1})A(xy^{-1})L(l^{-1})g(y) \\ &= R(xy^{-1})l^{-1}(xy)^{-1}f(xy^{-1})L(xy^{-1} \cdot l^{-1} \cdot (xy^{-1})^{-1})A(xy^{-1})g(y). \end{aligned}$$

Since, by 3) above, we have $R(l)\alpha \cdot L(l)\beta = \alpha \cdot \beta$ we can conclude that $F(ly) = f(xy^{-1})A(xy^{-1})g(y) = F(y)$.

This having been said, we can form the expression

$$(+) \qquad \int_H f(xy^{-1})A(xy^{-1})g(y) \cdot dh.$$

LEMMA 2. *We claim that $\mathcal{B}$ is a subalgebra of $C^*(G)$ and the product of $f, g \in \mathcal{B}$ is given by* (+).

In fact,

a) Let us write $F(x) \equiv \int_H f(xy^{-1})A(xy^{-1})g(y)\,dh$, and assume that $l \in L$. We claim that $F(lx) = R(l)F(x)$.

In fact,

$$\begin{aligned} F(lx) &= \int_H f(lxy^{-1})A(l)A(xy^{-1})g(y)\,dh \\ &= \int_H R(l)f(xy^{-1})L(l)R(l)A(xy^{-1})g(y)\,dh \\ &= \int_H f(xy^{-1})R(l)A(xy^{-1})g(y)\,dh \\ &= R(l)\left(\int_H f(xy^{-1})A(xy^{-1})g(y)\cdot dh\right) = R(l)F(x) \end{aligned}$$

where we have just made use of $R(l)(\alpha\cdot\beta) = \alpha R(l)\beta$ (cf. 4) above). We can thus conclude that $F(lx) = R(l)F(x)$.

b) We recall that

$$F(x) = \int_H f(xy^{-1})A(xy^{-1})g(y)\,dh$$

which clearly implies that $xL \mapsto \|F(x)\|$ is of compact support on $H = G/L$. In this manner, to complete the proof of Lemma 2 it suffices to establish the following.

c) Assume that F is as above, and that U is some unitary representation of G. In this case we have

$$\int_H U(F(x))U(x)\cdot dh = U(f)U(g).$$

In fact,

$$\begin{aligned} U(F(x)) &= \int_H U(f(xy^{-1}))U(A(xy^{-1})\cdot g(y))\cdot dh' \\ &= \int_H U(f(xy^{-1}))U(xy^{-1})U(g(y))U(y)\cdot dh' U(x^{-1}) \end{aligned}$$

which implies that

$$\begin{aligned} \int_H U(F(x))U(x)\cdot dh &= \int_H\left(\int_H U(f(xy^{-1})U(x^{-1}y)\,dh')U(g(y))U(y)\cdot dh\right. \\ &= U(f)U(g). \end{aligned}$$

Summing up, we have completed proving the following: $\mathcal{B}$ is a subalgebra of $C^*(G)$ and, if $f, g \in \mathcal{B}$, then

$$(fg)(x) = \int_H f(xy^{-1})A(xy^{-1})g(y)\,dh$$

which was the claim of Lemma 2.

LEMMA 3. *Assume that f is some element of $\mathcal{B}$. Then $f^* \in \mathcal{B}$ and $f^*(x) = A(x)((f(x^{-1}))^s)$.*

In fact, let us put $F(x) \equiv A(x)((f(x^{-1}))^s)$.

a) We note first that F lies in $\mathcal{B}$. This follows from the following series of relations:

1) If $l \in L$, then

$$\begin{aligned} F(lx) &= A(l)A(x)(f(x^{-1}l^{-1}))^s = A(l)A(x)(f(x^{-1}l^{-1}x \cdot x^{-1}))^s \\ &= A(l)A(x)(R(x^{-1}lx)f(x^{-1}))^s \\ &= A(l)A(x)L(x^{-1}l^{-1}x)(f(x^{-1}))^s \\ &= A(l)L(l^{-1})A(x)(f(x^{-1}))^s = R(A(x))(f(x^{-1}))^s = R(l)F(x) \end{aligned}$$

whence $F(lx) = R(l)F(x)$.

2) Evidently, $xL \mapsto \|F(x)\|$ is of a compact support on H.

We note that 1), 2) as above jointly imply that $F(x) \equiv A(x)((f(x^{-1}))^s)$ lies in $\mathcal{B}$, in which manner, to complete the demonstration of Lemma 3, it is enough to note

b) $U(F) = (U(f))^*$ for all U in $\operatorname{Rep}(G)$.

In fact,

$$U(F(x)) = U(A(x)(f(x^{-1}))^s) = U(x)(U(f(x^{-1}))U(x^{-1})$$

which implies

$$\int_H U(F(x))U(x)\,dh = \int_H U(x)(U(f(x^{-1}))^* \cdot dh$$

whence, by $x \mapsto x^{-1}$ we get

$$\int_H U(x^{-1})(U(f(x)))^* \cdot dh = \left(\int_H U(f(x))U(x)\,dh\right)^* = (U(f))^*$$

or $U(F(x)) = (U(f))^*$ as claimed above, completing the proof of Lemma 3.

In the following we continue to write $\mathcal{L} = C^*(L)$ but $\widehat{L}$ in place of $\widehat{\mathcal{L}}$.

We recall that $\widehat{L}/\widetilde{G}$ is countably separated (cf. Lemma 4 in 3.2). Let E be a fixed orbit. We assume that $\dim(E) > 0$. We recall that E is open in its closure (cf. [**Gl**], Theorem 1, p. 124, (3) $\Rightarrow$ (1) and (3) $\Rightarrow$ (6)). Given $\pi \in E$, fixed, the map $a\widetilde{G}_\pi \mapsto a\pi$ is a $\widetilde{G}$-equivariant homeomorphism. We conclude from this that spacewise E is a connected abelian Lie group, which carries a $\widetilde{G}$-invariant Radon measure ω.

Below we shall employ the following notations: $\mathfrak{H}$ = complex Hilbert space, $U(\mathfrak{H})$ = unitary group with the strong topology, $B(\mathfrak{H}) = \{$bounded operators with the strong topology$\}$, $C(\mathfrak{H}) = \{$compact operators with the uniform topology$\}$, $C_2(\mathfrak{H}) = \{$Hilbert-Schmidt operators$\}$. If $\dim(\mathfrak{H}) < +\infty$, $C(\mathfrak{H}) = B(\mathfrak{H})$.

LEMMA 4. *There is an open subset $O \subset E$, the complement of which is of ω-measure zero. There is a fixed Hilbert space $\mathfrak{H}$, a map $O \times L \to \mathcal{U}(\mathfrak{H})$, denoted $(\pi, l) \mapsto \rho_\pi(l)$, such that for each π fixed in O, $l \mapsto \rho_\pi(l)$ $(l \in L)$ is an irreducible representation of class π of L. The map $O \times L \to \mathcal{U}(\mathfrak{H})$ and the corresponding map $O \times \mathcal{L} \to B(\mathfrak{H})$ are strongly continuous. Given $f : O \to C(\mathfrak{H})$, continuous, tending to zero on the boundary, there is an element $w \in \mathcal{L}$ such that*

$$f(\pi) \equiv \rho_\pi(w) \qquad (\pi \in O).$$

In fact,

a) Since E is locally closed in $\widehat{L}$, there is F closed, O open in $\widehat{L}$ such that $E = F \cap O$.

We can assume that F and O are $\widetilde{G}$-invariant. In fact, we may suppose from the beginning that $F = \overline{E}$, and thus it is enough to replace O by $U_{a\in\widetilde{G}} aO$. We conclude from this that there are $\widetilde{G}$-invariant open sets O_k $(k = 1, 2)$ in $\widehat{L}$ such that $O_1 \supset O_2$ and $E = O_1 - O_2$.

b) We note that if O is $\widetilde{G}$-invariant and open in $\widehat{L}$, and I is the closed 2-sided ideal in $C^*(L)$ such that $O = \widehat{I}$, then I is $\widetilde{G}$-invariant. In fact, we have $\widehat{I} = \widehat{L} - F$ where F is $\widetilde{G}$-invariant and closed. We have $I = \ker(F)$. If $a \in \widetilde{G}$, then

$$aI = a\ker(F) = \ker(aF) = \ker(F) = I$$

and thus $aI = I$ $(a \in \widetilde{G})$.

We conclude from this that there are closed, 2-sided ideals in $\mathcal{L}$, both $\widetilde{G}$-invariant such that $\widehat{I_k} = O_k$ $(k = 1, 2)$. Let us put $D = I_1/I_2$; then $\widehat{D} = E$. The $\widetilde{G}$-action on D corresponds to the $\widetilde{G}$-action on E.

c) Next we recall the definition of C^*-algebras with continuous trace (cf. [**D***], 4.5, p. 92).

1) Suppose that A is a C^*-algebra. We denote by $\mathfrak{u}$ the set of all those elements x in A for which the function $\pi \mapsto \operatorname{Tr}(\pi(x)(\pi(x))^*)$ $(\pi \in \widehat{A})$ is finite and continuous. Then $\mathfrak{u}$ is a 2-sided ideal in A. This being so we say that A is of a continuous trace if and only if $\mathfrak{u}$ is dense in A. We note here the evident fact that if A is of a continuous trace and I is a closed 2-sided ideal in A, then I, too, is of a continuous trace.

2) We have the following fact. If A is a C^*-algebra, then it is of a continuous trace if and only if

(ι) $\widehat{A}$ is Hausdorff,

$(\iota\iota)$ given π_0 in $\widehat{A}$, there is p in A such that $\pi(p)$ is a projection of rank one for all π in a neighborhood of π_0.

This having been said we claim that D, as the end of b) above, is of a continuous trace. In fact,

1) We have $\widehat{D} = E$, which is Hausdorff, and thus D is postliminaire.

2) We recall (cf. [**D***], 4.5.5, p. 94) that there is $I \subset D$, a nonzero, closed 2-sided ideal that is of a continuous trace. Let $\pi_0 \in \widehat{D}$. Suppose that $a \in \widetilde{G}$, $\rho_0 \in \widehat{I}$ are such that $\pi_0 = a\rho_0$. Our assumption says that there is a neighborhood $U(\rho_0)$ of ρ_0 in $\widehat{I}$, and p in I such that $\rho(p)$ is a projection of rank one for all ρ in $U(\rho_0)$. Let us put $q = ap$. To complete our proof it is enough to note that $\pi(q)$ is a projection of rank one if $\pi \in a(U(\rho_0))$.

d) Before proceeding, we make some general observations. Suppose that T is a locally compact space with a countable base. Let $\mathfrak{H}$ be a separable Hilbert space such that $0 < \dim(\mathfrak{H}) \leq +\infty$. We denote by $S(T)$ the set of all continuous maps from T into $C(\mathfrak{H})$, which tend to zero at infinity. We note that $S(T)$ can be turned into a C^*-algebra by pointwise linear combinations, multiplication, taking adjoints and with sup norm. Our next objective will be to show that if O is open and contractible in E such that $\omega(E - O) = 0$, D' a closed, 2-sided ideal of D (cf. b) end) satisfying $\widehat{D'} = O$, then D' is of the form $S(O)$ with $\dim(\mathfrak{H}) \equiv \dim(\pi)$ $(\pi \in E)$. In fact,

1) Let us assume first that $\dim(\pi) \equiv +\infty$ $(\pi \in E)$. Here we can obtain the desired conclusion directly from the following result (cf. [**D***], 10.9.6, p. 219). Assume also that T is of a finite dimension and that $H^3(T,\mathbb{Z})$ is equal to zero. If A is a C^*-algebra of a continuous trace such that $\widehat{A} = T$, and that any irreducible representation of A is infinite-dimensional, then A is isomorphic to $S(T)$. Let us take now $A = D'$. D' is of a continuous trace, has no finite-dimensional representation and $\widehat{D}' = O$. Since O is contractible, we have $H^3(T,\mathbb{Z}) = 0$. Hence we conclude that D' is isomorphic to $S(O)$.

2) Here we assume that $\dim(\pi) < +\infty$. For more details about the following (cf. [**D***], 10.7, p. 205 and 10.9.2, p. 218). Let $\mathscr{H} = (H(t), \Gamma; t \in T)$ be a continuous field of Hilbert spaces. It gives rise canonically to a continuous field of C^*- algebras; the fiber over t is $C(H(t))$. The set of all continuous sections tending to zero at infinity gives rise, similarly as in the case of $S(T)$, to a C^*-algebra, $= A$ say, where A is of a continuous trace, and we have $\widehat{A} = T$.

Conversely, assume that A is of a continuous trace such that $H^0(\widehat{A},\mathbb{Z}) = 0$. Then it arises in the manner just described. Take now $A = D'$. Then all the above conditions are met. Since $\dim(H(t))$ $(t \in T)$ is finite, $\mathscr{H}$ is locally trivial. Hence, it is a Hilbert bundle which, since O is contractible, is also globally trivial. Thus again D' is of the form $S(O)$. In the following we shall assume that $\dim(\pi) \equiv +\infty$ $(\pi \in E)$ and leave to the reader the remaining cases.

e) We recall that $D = I_1/I_2$. Hence $D' = I/I_2$, where I is a closed ideal in I_1. Given $a \in D'$, we denote by $\pi \mapsto f(a,\pi)$ the map representing a in $S(O)$. Putting, for $a \in D'$ fixed, $\sigma_\pi(a) \equiv f(a,\pi)$ $(a \in D')$, lift σ_π to I, next extend the result first to I_1 (I is a closed 2-sided ideal in I_1) and next to $\mathcal{L} = C^*(L)$ (for this cf. [**D***], 2.10.4, p. 52). We denote the result by ρ_π $(\in \mathrm{Irr}(\mathcal{L}))$. We write also ρ_π for the corresponding representation of L.

f) We note that the map $\rho.(\cdot) : O \times L \to \mathcal{U}(\mathfrak{H})$ is continuous. In fact, let us write again R for the right-regular representation of L on $\mathcal{L}$. We note that R leaves invariant any closed 2-sided ideal in $\mathcal{L}$. If u, v are elements in I, then first $R(l^{-1})u \in I$ and $R(l^{-1})u \cdot v$ lies in Γ. We have $\rho_\pi(u)\rho_\pi(l)\rho_\pi(v) \equiv \rho_\pi(R(l^{-1})u \cdot v)$. By d) the right-hand side gives rise to a continuous map $O \times L \to C(\mathfrak{H})$. Let A, B be given, fixed in $C(\mathfrak{H})$. For a $\pi_0 \in O$ there is a neighborhood $U(\pi_0)$ of π_0, elements u, v in I, such that if $\pi \in U(\pi_0)$, then $\rho_\pi(u) \equiv A$, $\rho_\pi(v) \equiv B$, and $A\rho_\pi(l)B \equiv \rho_\pi(R(l^{-1})uv)$. In this manner it suffices to note that if U_n, U are in $\mathcal{U}(\mathfrak{H})$ such that $AU_nB \to AUB$ uniformly, for each choice, as above, of $A, B \in C(\mathfrak{H})$, then $U_n \to U$ strongly.

Summing up, we have completed proving that $(\pi, l) \mapsto \rho_\pi(l)$ is continuous from $O \times L$ to $\mathcal{U}(\mathfrak{H})$.

g) From here we finish our proof of Lemma 4 by noting that the map $O \times \mathcal{L} \to B(\mathfrak{H})$, defined by $(\pi, v) \mapsto \rho_\pi(v)$ $(\pi \in O, v \in \mathcal{L})$ is strongly continuous. In fact, suppose that $f \in \mathfrak{H}, \pi_0 \in O$ and $v \in \mathcal{L}$ are fixed. We have to show that $\|(\rho_\pi(v) - \rho_{\pi_0}(v_0)f\| \to 0$ if π and v tend respectively to π_0 and v_0. The square of the left-hand side is equal to

$$(\rho_\pi(v)f, \rho_\pi(v)f) - (\rho_\pi(v)f, \rho_{\pi_0}(v_0)f) - (\rho_{\pi_0}(v_0)f, \rho_\pi(v)) + (\rho_{\pi_0}(v_0)f, \rho_\pi(v_0)f).$$

Let P be a fixed finite-dimensional projection such that $Pf = f$. We can assume w in I to be such that $\rho_\pi(w) \equiv P$ if π is close to π_0.

Then the first term is equal to $(\rho_\pi(w^*v^* \cdot vuf, f))$, which tends to $(\rho_\pi(w^* \cdot v_0^*v_0w)f, f) \equiv (\rho_{\pi_0}(v_0)f, \rho_{\pi_0}(v)f)$ if $\pi \to \pi_0$ and $v \to v_0$. The remaining terms can be treated in a similar manner.

Given $A \in B(\mathfrak{H})$ we write $[A]$ for its Hilbert-Schmidt norm $(\leqq +\infty)$. We continue to write $H = G/L$, and dh for its Haar measure. We shall also need a smooth transversal H' modulo L; fixing one, we write again dh for the image of the latter on H'. We consider the following set: $\{A; A : G \times E \to C_2(\mathfrak{H})$ is weakly measurable such that

$$A(lx, \pi) \equiv A(x, \pi)(\rho_\pi(l))^* \quad \text{for all } x \in G, \pi \in E, l \in L$$

and

$$\int_{H\times E} [A(x,\pi)]^2 \, dh \, d\omega < +\infty\}.$$

We write $H(V)$ for the corresponding Hilbert space. For $a \in G$, fixed, consider the map $A(x, \pi) \mapsto A(xa, \pi)$. It gives rise to a continuous unitary representation of G on $H(V)$. This admits a simple group-theoretic interpretation. To see this, let us denote by m $(\leqq +\infty)$ the dimension of π $(\pi \in E)$. Putting

$$\Pi = \int_E^\oplus \pi \, d\omega$$

we have $V = m(\mathrm{ind}_{L\uparrow G} \Pi)$. Below, in particular, we shall show that $R(V)$ is semifinite. From this, using the Main Lemma (1.3), we could easily derive a new demonstration for the semifiniteness of the regular representation for a simply connected G, but we shall not pursue this idea and its ramification in the present work.

The map $(x, \pi) \mapsto \rho_\pi(f(x))$ is strongly continuous, and hence also weakly measurable on $G \times E$. Setting $A(x, \pi) \equiv \rho_\pi(f(x))$ we have $A(lx, \pi) \equiv A(x, \pi)(\rho_\pi(l))^*$ for any $l \in L, x \in G$ and $\pi \in E$.

We write

$$\|f\|_\omega^2 = \int_{H\times E} [\pi(f(x))]^2 \, dh \, d\omega \quad (\leqq +\infty).$$

Next let us consider the subset $\mathcal{B}' = \{f; f \in \mathcal{B}, \|f\|_\omega < +\infty\}$ of $\mathcal{B}$. We denote by ψ the map from $\mathcal{B}'$ into $H(V)$ defined by $\psi(f)(x, \pi) \equiv \rho_\pi(f(x))$ (f fixed in $\mathcal{B}'$; $x \in G, \pi \in E$).

LEMMA 5. *Unexplained notation as above. We claim that $\psi(\mathcal{B}')$ is dense in $H(V)$.*

In fact, let us write $\mathfrak{h} = L^2_\omega(E) \otimes C_2(\mathfrak{H})$; with the obvious identification $H(V) = L^2(H') \otimes \mathfrak{h}$. If, for some φ on H' and v in $\mathcal{L}$, we write $f = v \otimes \varphi \in \mathcal{B}$, then

$$\|f\|_\omega^2 = \|\varphi\|^2 \left(\int_E [\pi(v)]^2 \cdot d\omega \right)$$

and hence it suffices to recall that, by Lemma 4, $\mathfrak{h}$ is spanned by its elements of the form $\pi \mapsto \rho_\pi(v)$ $(\pi \in E, v \in I$, fixed).

We recall (cf. Lemma 3) that for $v \in \mathcal{L}$, v^s stands for its adjoint and if $f \in \mathcal{B}$, then $f^*(x) \equiv \Delta(x)((f(x^{-1}))^s)$.

LEMMA 6. *Let f be some element of $\mathcal{B}$. Then we have*

$$\|f^*\|_\omega = \|f\|_\omega \qquad (\leqq +\infty).$$

In fact, we have

$$[\pi(f^*(x))] = [\pi(A(x)((f(x)^{-1}))^s)] = [(x^{-1}\pi)(f(x^{-1}))]$$

from which, by the G-invariance of ω, we conclude that

$$\begin{aligned}\|f^*\|_\omega^2 &= \int_{H\times E} [(x^{-1}\pi)(f(x^{-1}))]^2 \cdot dh \cdot d\omega \\ &= \int_{H\times E} [\pi(f(x))]^2 \cdot dh \cdot d\omega = \|f\|_\omega^2\end{aligned}$$

and thus

$$\|f^*\|_\omega^2 = \|f\|_\omega^2 \qquad (\leqq +\infty).$$

We note that on $C^*(G)$ there is a canonical conjugation extending $(f \mapsto \overline{f})$ $(f \in \mathcal{L}(G))$. We define for $f \in C^*(G) : f^t \equiv (\overline{f})^*$. If $f \in \mathcal{L}(G)$, then $f^t(x) \equiv f(x^{-1})/\Delta(x)$.

LEMMA 7. *Assume that f and g are elements of $\mathcal{B}'$ and $\mathcal{B}$ resp. Then $f \cdot g \in \mathcal{B}'$ and $\psi(fg) = V(g^t)\psi(f)$.*

In fact,

a) We observe that if K is compact in H, then there is a positive constant C, depending on K only, such that if f in $\mathcal{B}'$ and g in $\mathcal{B}$ satisfy $\operatorname{supp}(f), \operatorname{supp}(g) \subset K$, then we have

$$\|fg\|_\omega \leqq C(\|g\|_\infty)\|f\|_\omega.$$

In fact, if A and B are bounded operators, then $[AB] \leqq [A]\|B\|$. We recall that

$$(fg)(x) = \int_H f(xy^{-1})A(xy^{-1})g(y) \cdot dh.$$

Hence we can conclude that

$$\begin{aligned}[\pi(fg)(x)] &\leq \int_H [f(xy^{-1})]\|g(y)\| \cdot dh \\ &\leqq \|g\|_\infty \int_H [\pi(fg)] \cdot dh.\end{aligned}$$

Putting $k = (h(\operatorname{supp}(f)))^{1/2}$ we obtain

$$\pi[(fg)(x)] \leqq k\|g\|_\infty \left(\int_H [\pi(fg)]^2\, dh\right)^{1/2}.$$

Observing that $\operatorname{supp}(fg) \subseteqq K^2$ we infer that

$$\begin{aligned}\|fg\|_\omega^2 &= \int_E \left(\int_{K^2} [\pi(fg)(x)]^2 \cdot dh\right) d\omega \\ &\leqq k^2 h(K^2)(\|g\|_\infty)^2\|f\|_\omega^2\end{aligned}$$

whence finally, if $C = k(h(K^2))^{1/2}$, then

$$\|f \cdot g\|_\omega \leq C\|g\|_\infty \cdot \|f\|_\omega.$$

b) We recall that if $f \in C^*(G)$, then $f^t = (\overline{f})^*$. Assume that $g \in \mathcal{B}_0$ and $f \in \mathcal{B}'$. We have evidently

$$(fg)(x) \equiv \int_G g^t(a) f(xa)\, da$$

from where we conclude that $fg \in \mathcal{B}'$ and $\psi(fg) = V(g^t)\psi(f)$.

c) Suppose that $g \in \mathcal{B}$, and $f \in \mathcal{B}'$. Assume that $(g_n) \subset \mathcal{B}_0$ is such that $g_n \to g$ in $\mathcal{B}$.

1) We recall that by Lemma 1, $g_n \to g$ in $C^*(G)$ whence we conclude that $g_n^t \to g^t$ in $C^*(G)$. We have by b): $\psi(fg_n) = V(g_n^t)\psi(f)$, from where we infer that $\lim_{n\to+\infty} \psi(fg_n) = V(g^t)\psi(f)$.

2) To obtain the desired conclusion we note that $\lim_{n\to+\infty} \psi(fg_n) = \psi(fg)$.

In fact, our assumptions imply that there is a fixed compact set in H containing the supports of $(g_n), g$ and f. Using this, we infer by a) that there is a constant $C > 0$, not depending on g_n, g and f such that

$$\|\psi(fg_n) - \psi(fg)\|_\omega = \|f(g - g_n)\|_\omega \leqq C\|g - g_n\|_\infty \cdot \|f\|_\omega$$

whence $\lim_{n\to+\infty} \psi(fg_n) = \psi(fg)$ as claimed in 2) above.

Since $\psi(fg) = \lim_{n\to+\infty} \psi(fg_n) = \psi(g^t)\psi(f)$ we conclude finally that if f and g are elements of $\mathcal{B}'$ and $\mathcal{B}$ respectively, then $f \cdot g \in \mathcal{B}'$ and $\psi(fg) = V(g^t)\psi(f)$ as claimed in Lemma 7.

LEMMA 8. *There is a sequence (g_n) in $\mathcal{B}'$ such that we have for any $f \in \mathcal{B}'$: $\psi(fg_n) \to \psi(f)$ in $H(V)$.*

In fact,

a) We recall that for $v \in \mathcal{L}$, v^s stands for its conjugate. We define for $v \in \mathcal{L}$: $v^r \in (\overline{v})^s$. We observe that for any $f \in \mathcal{B}$, we have

$$f^t(x) \equiv A(x)((f(x^{-1}))^r).$$

In fact, to see this we can assume that f is in $\mathcal{B}_0$. We have

$$f^t(x) \equiv (l \mapsto f(lx)^{-1})/\Delta(x).$$

But $f((lx)^r) \equiv ((x^{-1}l^{-1}) \equiv f((x^{-1}lx)^{-1} \cdot x^{-1})$ which clearly implies our assertion.

b) We note next that given $v \in \mathcal{L}$ such that

$$\int_E [\pi(\overline{v})]^2 \cdot d\omega < +\infty$$

and putting, for $\varphi \in \mathcal{L}(H') : f = \varphi \otimes v \in \mathcal{B}$, we have $f^t \in \mathcal{B}'$. To see this, we note that for any $f \in H'$, we have

$$\begin{aligned}[\pi(f^t(h))] &= [\pi(A(h))(f(h^{-1}))^r)] \\ &= [(h^{-1}\pi)(v)][\varphi(h^{-1})]\end{aligned}$$

whence we obtain

$$\begin{aligned}\|f^t\|_\omega^2 &= \int_{H'} |\varphi(h)|^2 \left(\int_E [h^{-1}\pi(\overline{v})]^2 \cdot d\omega\right) dh \\ &= \|\varphi\|^2 \left(\int_E [\pi(\overline{v})]^2\, d\omega\right) < +\infty.\end{aligned}$$

Summing up, we have proved that if $v \in \mathcal{L}$ is such that

$$\int_E [\pi(\overline{v})]^2 \cdot d\omega < +\infty$$

and $\varphi \in \mathcal{L}(H')$ and $f = v \otimes \varphi \in \mathcal{B}$, then f^t is in $\mathcal{B}'$.

c) We note that if φ is in $\mathcal{L}(H')$, $V(\varphi) = \int_{H'} \varphi(h)V(h)\,dh$ and $v \in \mathcal{L}$, then

$$V(v \otimes \varphi) = V(v)V(\varphi).$$

In fact, we have

$$\begin{aligned} V(v \otimes \varphi) &= \int_{H'} V(v \otimes \varphi)(h)) \cdot V(h)\,dh \\ &= V(v) \int_{H'} \varphi(h)V(h)\,dh = V(v)V(\varphi). \end{aligned}$$

d) Let us make now the following assumptions. Let $(v_n) \subset \mathcal{L}$ be such that

$$\int_E [\pi(\overline{v}_n)]^2\, d\omega < +\infty$$

and $V(v_n)$ tends to the unity strongly.

Assume also that $(\varphi_n) \subset \mathcal{L}(H')$ is such that $V(\varphi_n) \to I$ strongly. Putting $g_n = (v_n \otimes \varphi_n)^t$ we claim that (g_n) is as stated in Lemma 8. In fact, by b), $(g_n) \subset \mathcal{B}'$. Let f be some element in $\mathcal{B}'$. Lemma 7 implies that $\psi(fg_n) = V(v_n \otimes \varphi_n)\psi(f)$ and by c), $V(v_n \otimes \varphi_n) = V(v_n)V(\varphi_n)$ tends to the unity strongly. From here, to complete the proof of Lemma 8, we need only

e) There is a sequence (v_n) in $\mathcal{L}$ such that

$$\int_E [\pi(\overline{v}_n)]^2 \cdot d\omega < +\infty$$

and $V(v_n)$ tends to the unity strongly. To this end we recall that V is quasi-equivalent to $\mathrm{ind}_{L\uparrow G}\,\Pi$ and Π is G-invariant. Hence $V|L$ is quasi-equivalent to Π. Therefore it suffices to arrange the choice of (v_n) such that $\Pi(v_n) \to$ unity strongly. We note that, evidently, the $\widetilde{G}$ action on $\widehat{L}$ commutes with the conjugation $\pi \to \overline{\pi}$ on $\widehat{L}$. Thus E and $\overline{E}$ are both $\widetilde{G}$-orbits, and we can distinguish two cases:

A) $E = \overline{E}$,

B) $E \cap \overline{E} = \varnothing$.

Unexplained notation below will be as in Lemma 2.

Case A) We have evidently that if $v \in \mathcal{L}$, then

$$\int_E [\pi(v)]^2 \cdot d\omega = \int_E [\pi(\overline{v})]^2 \cdot d\omega \qquad (\leqq +\infty).$$

We construct (v_n) as follows. Let (O_n) be open, increasing, and of compact closure in O. We assume that $\bigcup_n O_n = O$. For any $n = 1, 2,$ we choose f_n such that $0 \leqq f_n \leqq 1$, $f_n|O_n \equiv 1$ and $\mathrm{supp}(f_n)$ is compact. Take $(v_n) \subset \mathcal{L}$ such that $\rho_\pi(v_n) \equiv f_n P_n$, where (P_n) are projections such that $P_1 < P_2 < \cdots$, $\dim(P_n) \equiv n$ $(n = 1, 2, \dots)$. We have $\int_E [\pi(\overline{v}_n)]^2 \cdot d\omega < +\infty$ and $(\Pi(v_n))$ and thus also $(V(v_n))$ tend to the unity strongly.

Case B) Here we assume that $E \cap \overline{E} = \varnothing$. Through a repetition of the reasoning in Lemma 2, we conclude that there is a subquotient D' of $\mathcal{L}$, such that D' is of a continuous trace, satisfying $\widehat{D}' = O \cup \overline{O}$. Thus we can select (v_n) in $\mathcal{L}$ such that

$$\int_E [\pi(\overline{v}_n)]^2 \cdot d\omega < +\infty$$

and $(\rho_\pi(v_n))$ is as in Case A).

Before proceeding we recall the following facts (cf. [**D vN**], p. 70). A Hilbert algebra A is an algebra over the complex field, endowed with a pre-Hilbert space structure and an involutive antiautomorphism $x \mapsto x^s$ $(x \in A)$ such that:

If x, y, z are in A, then we have

(1) $(x, x) = (x^s, x^s)$,

(2) $(xz, y) = (x, yz^s)$,

(3) for any $x \in A$, fixed, $y \mapsto yx$ is continuous,

(4) $\{xy; x, y \in A\}$ is dense in A.

We denote by $H(A)$ the completion of A. Given $x \in A$, (3) implies the existence of an operation V_x in $B(H(A))$ well determined by $V_x y \equiv yx$ $(y \in A)$. We shall denote by $R(A)$ the von Neumann algebra generated by $(V_x; x \in A)$.

Next we note that $R(A)$ is semifinite. In fact, there is a faithful, normal trace ϕ on $R(A)$ satisfying

$$(x, x) = \phi(V_x \cdot V_x^*) \qquad (x \in A).$$

LEMMA 9. *Unexplained notation as above. We claim that there is a Hilbert algebra structure $\mathcal{E}$ on $\psi(\mathcal{B}')$ compatible with the natural operations such that $R(\mathcal{E}) = R(V)$.*

In fact,

a) We set $I = \ker(\psi)$ and claim that I is a 2-sided ideal in $\mathcal{B}'$. In fact,

1) If f is in I and g is in $\psi(\mathcal{B}')$, then $\psi(fg) = V(g^t)\psi(f) = 0$ (by $\psi(f) = 0$).

2) We have, by Lemma 6, that $I^* = I$.

3) We have $gf = (f^* \cdot g^*)^*$ and thus by 1) $\psi(gf) = 0$. We conclude that we can form $\mathcal{E} = \mathcal{B}'/I \leftarrow \mathcal{B}' : \psi$. We define, if $x \in \mathcal{E}$ and $x = \psi(f) : x^s = \psi(f^*)$. If $x, y \in \mathcal{E}$ are such that $x = \psi(f), y = \psi(g)$ then $(x, y) = (\psi(f), \psi(g))$.

b) We claim that the conditions (1)–(4) are satisfied. In fact,

Condition (1). We recall that by Lemma 6, $\|f\|_\omega = \|f^*\|_\omega$. If $x \in \mathcal{E}$, $x = \psi(f)$ then, by definition, $x^s = \psi(f^*)$ and hence $(x^s, x^s) = \|f^*\|_\omega^2 = \|f\|_\omega^2 = (x, x)$ in which fashion $(x, x) = (x^s, x^s)$.

Condition (2). Assume that x, y, z are given in $\mathcal{E}$ such that (say) $x = \psi(f), y = \psi(g)$, $z = \psi(h)$. We claim that $(xz, y) = (x, yz^s)$. We have, in fact,

$$\begin{aligned}(xz, y) &= (\psi(f(h), \psi(g)) = (V(h^t)\psi(f), \psi(g)) \\ &= (\psi(f), (V(h^t))^* \cdot \psi(g)).\end{aligned}$$

But $(V(h^t))^* = (V(h^*))^t$ and thus

$$\begin{aligned}(xz, y) &= (\psi(f), (V(h^*))^t \psi(g)) \\ &= (\psi(f), \psi(gh^*)) = (x, yz^s)\end{aligned}$$

whence

$$(xz, y) = (x, yz^s).$$

Condition (3). It is enough to note that if $x \in \mathcal{E}$ is fixed, $x = \psi(f)$, say $(f \in \mathcal{B}')$, and $y = \psi(g)$ (g arbitrary in $\mathcal{B}'$) then $yx = \psi(gf) = V(f^t)y$ and thus $y \mapsto yx$ is continuous.

Condition (4). This is implied by Lemma 8.

c) We conclude the proof of Lemma 9 by showing that $R(\mathcal{E}) = R(V)$. In fact,

1) Evidently, given a fixed x in $\mathcal{E}$, there is a bounded operator U_x on $H(V)$ such that $U_x y \equiv xy$ $(y \in \mathcal{E})$. We set

$$R(\mathcal{E}) = \{V_x; x \in \mathcal{E}\}''$$
$$L(\mathcal{E}) = \{U_x; x \in \mathcal{E}\}''$$

and recall that $(L(\mathcal{E}))' = R(\mathcal{E})$.

2) We note that $R(\mathcal{E}) \subseteqq R(V)$. To see this, let f and g be elements of $\mathcal{B}'$ such that $x = \psi(f), y = \psi(g)$. Then

$$V_x y = \psi(gf) = V(f^t)y.$$

3) We note next that if $x \in \mathcal{E}$ and $a \in G$, then U_x and $V(a)$ commute. To show this, we observe first that putting $f_a(x) = f(xa)$ we have $V(a)\psi(f) \equiv \psi(f_a)$. This being so, it suffices to prove that for $f, g \in \mathcal{B}'$ we have $(gf)_a = gf_a$. To this end, we recall that

$$(gf)(x) = \int_H g(xy^{-1})A(xy^{-1})f(y) \cdot dh$$

from where we get

$$\begin{aligned}(gf)_a(x) &= \int_H g(xay^{-1})A(xay^{-1})f(y)\,dh \\ &= \int_H g(xy^{-1})A(xy^{-1})f(ya)\,dh = g \cdot f_a(x)\end{aligned}$$

implying $(gf)_a = g \cdot f_a$.

4) Summing up, 3) implies that $L(\mathcal{E}) \subseteqq (R(V))'$, whence $R(\mathcal{E}) = (L(\mathcal{E}))' \supseteqq R(V) \supseteqq R(\mathcal{E})$ from which we conclude that $R(\mathcal{E}) = R(V)$, completing our proof of Lemma 9.

3.5. Existence of normal representations with given kernels

We recall that, given a locally compact group G, a unitary representation is called normal if it generates a semifinite factor such that for some f in $C^*(G)$, $f^* \cdot f$ admits a positive finite trace. We note also that the kernel of a factor representation is a primitive ideal (cf. [**Dal**], p. 100). Next we assume again that G is a connected and simply connected Lie group. The purpose of this section will be to show that, given a primitive ideal J of $C^*(G)$, there is a normal representation of G, the kernel of which is equal to J. We recall that the principal objective of Chapter III will be to establish that, in addition, two normal representations with the same kernel are quasi-equivalent. For unexplained notation below we refer to Lemmas 5 and 6 in Section 3.2.

LEMMA 1. *Let J be a fixed element of* Prim(G). *If O is a Σ-orbit in the corresponding F, then there is a positive G-invariant Radon measure on O. It is well-determined up to a positive constant factor.*

In fact, let ρ be some element of O. Then O is the relative closure, in F, of the G-orbit of ρ. We recall that $\iota : N/N_\rho \to F$ is an N-equivariant homeomorphism. If $\mathcal{A}$ is the closure of the image of G in N/N_ρ, then so is the map $\iota|\mathcal{A}$ onto O. Hence, to arrive at our goal, it is enough to take a G-invariant Radon measure on $\mathcal{A}$, and form its image on O, and similarly in the opposite direction.

We observe that there is a Borel cross section $\sigma : F \to \mathrm{Irr}(K)$. In fact, let now ρ be a concrete representation. If $C : N/N_\rho \to N$ is a Borel cross section, it is enough to take $\sigma(\zeta) \equiv c(\zeta)\rho$ $(\zeta \in F)$.

Below we write $\rho(\zeta)$ in place of $\sigma(\zeta)$. Also, let $\mathfrak{G}$ be $C^*(G)$, $\mathfrak{G}^+$ the set of all positive elements in $\mathfrak{G}$, and $\mathcal{B}^+ = \mathcal{B} \cap \mathfrak{G}^+$.

LEMMA 2. *With the previous notation, let μ be a G-invariant positive Radon measure on O.*

We put

$$T = \mathrm{ind}_{K\uparrow G}\left(\int_O \rho(\zeta)\cdot d\mu(\zeta)\right)$$

and claim that T is a normal representation of G such that $\ker(T) = J$.

We have already shown in Section 3.3 that T is a factor representation with the kernel J. Therefore it suffices to show that T is normal.

a) Let ω be a Radon measure on E $(= E(J))$ that is $\widetilde{G}$-invariant. Let $\pi : E \to \mathrm{Irr}(L)$ be a Borel cross section. We set $A = N/N_\rho$ and $\iota(aN_\rho) = a\rho$. For a Haar measure τ' on A let us put $\tau = \iota_*(\tau')$. Writing

$$Y = \int_F^\oplus \rho(\zeta)\cdot d\tau(\zeta)$$

we claim that $Y = \mathrm{ind}_{L\uparrow K}\Pi$.

In fact, define $P : F \to E$ by $P(\rho) = \rho|L$ $(\rho \in F)$ and take the corresponding disintegration of measures

$$\tau = \int_E \sigma_\eta \cdot d\omega(\eta).$$

This can be explained in the following manner: Let $\mathcal{F}$ be the annihilator of L in $\mathrm{X}(K)$. Then σ_η is $\mathcal{F}$-invariant on F and is carried by $P^{-1}(\eta)$.

Let us put for $\eta \in E : Y_\eta = \int_F^\oplus \rho(\zeta)\, d\sigma_\eta(\zeta)$. By [**Mackann I**], Th. 10.2, p. 123 and [**Mackann II**], Th. 2.1, p. 204 we conclude that the map $\eta \mapsto Y_\eta$ is ω-measurable and we have $Y = \int_E^\oplus Y_\eta\, d\omega(\eta)$. But (cf. [**Mack Jer**], Lemma 1, p. 325) we have also $Y_\eta = \mathrm{ind}_{L\uparrow K}\pi(\eta)$. Therefore

$$\begin{aligned} Y &= \int_E^\oplus Y_\eta \cdot d\omega(\eta) = \int_E^\oplus \left(\mathrm{ind}_{L\uparrow K}\pi(\eta)\right) d\omega(\eta) \\ &= \mathrm{ind}_{L\uparrow K} \int_E^\oplus \pi(\eta)\, d\omega(\eta) = \mathrm{ind}_{L\uparrow K}\Pi \end{aligned}$$

whence

$$Y = \mathrm{ind}_{L\uparrow K}\Pi$$

as defined above.

b) We recall that F can be viewed as underlying to a connected abelian group. The portion of Σ in F arises as the fibration corresponding to a closed connected subgroup. We write $\Lambda = F/\Sigma$ and, given $\lambda \in \Lambda$, we denote by O_λ the corresponding

Σ-orbit. By what we have just said, there is a Radon measure ν on Λ, and for each $\lambda \in \Lambda$, a G-invariant Radon measure carried by O_λ in F such that

$$\tau = \int_\Lambda \tau_\lambda \cdot d\nu(\lambda).$$

We conclude from here that, putting $Z(\lambda) = \int_F^\oplus \rho(\zeta)\, d\tau_\lambda(\zeta)$, then by a) we have

$$Y = \int_F^\oplus \rho(\zeta)\, d\tau(\zeta) = \int_\Lambda^\oplus Z(\lambda)\, d\nu(\lambda).$$

Let us write $U = \operatorname{ind}_{L\uparrow G} \Pi$. Setting $U(\lambda) = \operatorname{ind}_{K\uparrow G} Z(\lambda)$ we note that $U(\lambda)$ is a factor representation of G. We put $\ker(U(\lambda)) = J_\lambda \in \operatorname{Prim}(G)$. Evidently if $J_\lambda = J_{\lambda'}$, then $\lambda = \lambda'$. This being said, we claim that

$$U = \int_\Lambda^\oplus U(\lambda) \cdot d\nu(\lambda).$$

We have in fact

$$\begin{aligned} U &= \operatorname{ind}_{L\uparrow G} \Pi = \operatorname{ind}_{K\uparrow G}\Big(\operatorname{ind}_{L\uparrow K} \Pi\Big) = \operatorname{ind}_{K\uparrow G} Y \\ &= \operatorname{ind}_{K\uparrow G}\Big(\int_\Lambda^\oplus Z(\lambda)\cdot d\nu(\lambda)\Big) = \int_\Lambda \Big(\operatorname{ind}_{K\uparrow G} Z(\lambda)\Big)\cdot d\nu(\lambda) \\ &= \int_\Lambda^\oplus U(\lambda)\cdot d\nu(\lambda) \end{aligned}$$

and hence

$$U = \int_\Lambda^\oplus U(\lambda)\cdot d\nu(\lambda) \tag{+}$$

as claimed above.

c) We note that (+) is central. To prove this, we shall make use of the following proposition, itself a special case of a more general result due to E. G. Effros (cf. [**E**], Theorem 1.10, p. 91). Let (Z, α) be a standard measure space, and $\zeta \mapsto T(\zeta)$ an α-measurable field on Z. We assume that $\ker(T(\zeta_1)) = \ker(T(\zeta_2))$ implies $\zeta_1 = \zeta_2$. Then the integral $\int_Z^\oplus T(\zeta)\cdot d\alpha(\zeta)$ is central. Since, as noted before, $\ker(U(\lambda_1)) = \ker(U(\lambda_2))$ implies $\lambda_1 = \lambda_2$, we have proved that (+) is central.

d) Below we shall denote by $\operatorname{Facn}(G)$ the totality of all factor normal representations of G. This being so our next claim is that if for some λ_0 in Λ, $U(\lambda_0) \in \operatorname{Facn}(G)$, then $U(\lambda)$ is in $\operatorname{Facn}(G)$ for all λ in Λ. In fact

1) Let us put $\widetilde{N} = \widetilde{G} \times \mathrm{X}(G)$. If $a = (b, \chi) \in \widetilde{N}$ and T is some representation of G, we write $(aT)(g) \equiv \chi(g)T(b^{-1}gb)$. In this fashion $\widetilde{N}$ acts on the left on $\operatorname{Rep}(G)$. The same is true for $N = \widetilde{G} \times \mathcal{F}$, where $\mathcal{F}$ is the annihilator of L in $\mathrm{X}(K)$.

2) We denote by $t : \mathrm{X}(G) \to \mathcal{F}$ the restriction to K, and if $a = (b, \chi)$ is some element in $\widetilde{N}$, then $j(a) = (b, t(\chi))$. We note, incidentally, that N acts transitively on $\Lambda = F/\Sigma$.

3) This having been said, we claim that if $a \in \widetilde{N}$ and $\lambda \in \Lambda$, then

$$aU(\lambda) = U(j(a)\lambda).$$

We write, as in b) above, $Z(\lambda) = \int_F^\oplus \rho(\zeta)\cdot d\tau_\lambda(\zeta)$ and $U(\lambda) = \operatorname{ind}_{K\uparrow G} Z(\lambda)$.

If $c = j(a)$ and $c\lambda = \kappa$, then c maps O_λ onto O_κ, and $d\tau_\lambda$ is carried into a multiple of $d\tau_\kappa$. This being so,

$$\begin{aligned} cZ(\lambda) &= \int_F^\oplus c(\rho(\zeta)) \cdot d\tau_\lambda(\zeta) \\ &= \int_F^\oplus \rho(c\zeta)\, d\tau_\lambda = \int_F^\oplus \rho(\zeta) \cdot d\tau_\kappa(\zeta) = Z(c\lambda) \end{aligned}$$

whence $cZ(\lambda) = Z(c\lambda)$ and thus also

$$aU(\lambda) = \mathrm{ind}_{K\uparrow G}\, Z(c\lambda) = U(j(a)\lambda)$$

and hence $aU(\lambda) = U(j(a)\lambda)$.

4) Assume that, for some λ_0 fixed, $U(\lambda_0) \in \mathrm{Facn}$.

Let $\lambda \in \Lambda$. There is $a \in \widetilde{N}$ such that $j(a)\lambda_0 = \lambda$ and thus also $aU(\lambda_0) = U(\lambda)$.

By Lemma 2 in Section 3.2, the action of $\widetilde{N}$ in $\mathrm{Rep}(G)$ is contragredient to an automorphism of $\mathfrak{G}$. So $U(\lambda) \in \mathrm{Facn}(G)$.

e) For more details and notation for what follows below cf. Lemma 9 in Section 3.4.

Reminder:

$$V = m\Big(\mathrm{ind}_{L\uparrow G}\,\Pi\Big) = mU \qquad (0 < m \leqq +\infty).$$

By loc. cit., there is a unitary algebra $\mathcal{E}$ such that $R(\mathcal{E}) = R(V)$ and $V(\mathcal{B}) \supset \{V_x; x \in \mathcal{E}\}$.

Let ϕ be the natural trace on $[R(\mathcal{E})]^+$, so that $\phi(V_x^* \cdot V_x) = (x, x)$.

We conclude that V is a representation with trace (to be denoted again by ϕ).

Since $U = \int_\Lambda^\oplus U(\lambda) \cdot d\nu(\lambda)$ is central, by [**D vN**], Théorème 2, (ii), p. 206, we conclude that there corresponds to it a decomposition

$$\phi = \int_\Lambda^\oplus \phi_\lambda \cdot d\nu(\lambda)$$

where ϕ_λ is faithful, normal, and semifinite on $[R(U(\lambda))]^+$. Assume that A is a positive element of $(R(U))^+$ such that $A = \int_\Lambda^\oplus A(\lambda)\, d\nu(\lambda)$. Then we have

$$\phi(A) = \int_\Lambda \phi_\lambda(A(\lambda)) \cdot d\nu(\lambda) \qquad (\leqq +\infty).$$

Let $a \in \mathcal{B}^+$ be such that $0 < \phi(U(a)) < +\infty$ (e.g., take $b \in \mathcal{B}'$ such that $U(b^t \cdot b) \neq 0$ and then set $a = b^*(b^t)^*$).

We conclude that there is $\delta \in \Lambda$ such that

$$0 < \phi_\delta(U(\delta)(a)) < +\infty.$$

Summing up all, we have obtained the following

PROPOSITION. *Let J be a given primitive ideal of G. Then $\mathfrak{t}(J) = \{\rho : \rho \in A, J(\rho) = J\}$ is locally compact, separable, carrying a positive G-invariant Radon measure $d\mu$ such that*

$$\mathrm{ind}_{K\uparrow G}\left(\int_{\mathfrak{t}(J)}^\oplus \rho\, d\mu\right)$$

is normal, with kernel equal to J.

3.6. The type I case

We propose to deal separately with the case when at least one of the factors involved is of type I. We write $\mathrm{Irrn}(G)$ for the collection of all irreducible, normal representations.

LEMMA 1. *We consider the following two properties bearing on the primitive ideal J.*

(1) *If $V \in \mathrm{Fac}(G)$ is such that $\ker(V) = J$, then $V = (I)$.*

(2) *There is an irreducible normal representation with the kernel J.*

We claim that these two properties are equivalent.

In fact, let us assume first that we have an irreducible normal representation W such that $\ker(W) = J$. Let V be a factor representation with the kernel J. We are to show that V is of type I. We undertake a disintegration, over a standard measure space, of V into irreducibles

$$V = \int_X^{\oplus} V(x)\, d\mu(x).$$

Then we have (cf. [**Dal**], p. 100), that for all $v \in C^*(G)$ and almost all x, $\|V(v)\| \equiv \|V(x)(v)\|$.

Thus $\ker(V(x)) \equiv J = \ker(W)$. Therefore, for almost all x, $V(x)$ is equivalent to W whence V itself is a multiple of W. In this fashion V is of type I. Conversely, assume (1). By the proposition of 3.5, there exists a normal factor representation U with the kernel J. Then U is of type I, and so a multiple of an irreducible normal representation, which concludes the proof of Lemma 1.

This being so, we introduce the following

DEFINITION. We say that a primitive ideal J is of type I, if it satisfies one of the two equivalent properties of Lemma 1.

This being so, we have the following

LEMMA 2. *Suppose that V_1, V_2 are normal factor representations with the same kernel of type* I. *Then they are quasi-equivalent.*

In fact, let W be an irreducible normal representation with the given kernel. Then, by what we saw in the proof of Lemma 1, we have $V_1 \approx W$ and $V_2 \approx W$ and thus $V_1 \approx V_2$.

3.7. The non-type I case

We consider again a connected and simply connected Lie group G. We fix a primitive ideal J and consider the corresponding $\widetilde{G}$ orbit $E = E(J)$. Let π be a fixed element in E. It admits a projective extension π^e to G_π, giving rise to a continuous cocycle α. We can and shall assume that

$$\alpha(g_1 l_1, g_2 l_2) = \alpha(g_1, g_2) \qquad (g_1, g_2 \in G_\pi; l_1, l_2 \in L).$$

We form the corresponding group extension

$$G_\pi^e = \{(a,u); a \in G_\pi, u \in T \text{ such that } (a,u)(b,v) = (ab, \alpha(a,b)uv)\}.$$

As before, we write also L for $\{(l,1); l \in L\}$; L is invariant in G^e_π. Below we shall write

$$\Gamma = M_\pi \ (= G^e_\pi/L \leftarrow G^e_\pi : \phi).$$

If H is a subset of a group, we write $C(H)$ for its centralizer; that is, $C(H) = \{\gamma; \gamma \in \Gamma, \gamma h = h\gamma \ (h \in H)\}$. We recall that for $\Gamma = M_\pi$ we have $\Gamma_1 = C(\Gamma_0), U = C(\Gamma_0^c)$, where $\Gamma_0^c = (\Gamma_0)^c$ and $U = \Gamma_1 \cdot \Gamma_0$. Below H will stand for a maximal abelian subgroup of Γ_1. Since the elements of H and Γ_0 commute, $\Gamma_0 H$ yields a subgroup of Γ; we shall denote it by $\mathcal{E}$.

LEMMA 1. *Unexplained notation as above. We claim that $\mathcal{E}$ is the centralizer of H and that $\mathcal{E}$ is of the form $\Gamma_0 \times \Delta$, where Δ is free abelian.*

To prove all this we start by noting that $\mathcal{E} \subseteqq C(H)$. This is so since $\mathcal{E} = \Gamma_0 \cdot H$, and that Γ_0 commutes with H by $H \subseteqq C(\Gamma_0)$; also, H is abelian. We show next that $C(H) \subseteqq \mathcal{E}$. In fact, if γ is an element of $C(H)$, then it lies also in $C(\Gamma_0^c) = U$. Since $U = \Gamma_0 \cdot \Gamma_1$ we can write $\gamma = \gamma_1 \cdot \gamma_0$ ($\gamma_1 \in \Gamma_1$ and $\gamma_0 \in \Gamma_0$). Thus γ_1 commutes with H and hence it is contained in H. Hence γ is in $H\Gamma_0 = \mathcal{E}$, implying that $C(H) \subseteqq \mathcal{E}$ and thus also $C(H) = \mathcal{E}$. We show next that $\mathcal{E} = \Gamma_0 \times \Delta$. Since $H_0 = H \cap \Gamma_0$ it is enough to prove that $H = H_0 \times \Delta$.

Since

$$H/H_0 = H/(H \cap \Gamma_0) = \Gamma_0 H/\Gamma_0,$$

H/H_0 is free abelian, and thus H is of the form $H = H_0 \times \Delta$. Hence finally $\mathcal{E} = \Gamma_0 \times \Delta$, where Δ is free abelian, proving our lemma.

LEMMA 2. *We claim that $\mathcal{E}$ is of type* I.

In fact, by what we have just seen, it suffices to note that Γ_0 is a connected nilpotent Lie group and Δ is free abelian.

Let T be the canonical torus in G^e_π. We define $D^e = \phi^{-1}(\mathcal{E})$, $D = D^e/T$. We recall that $K = \phi^{-1}(U^c)/T$. In fact, $U^c \subset H \subset \mathcal{E} = H\Gamma_0 \subset U$, and $K = \phi^{-1}(U^e)/\Gamma$, $U = \phi^{-1}(U)/T$. We recall that J is a fixed element of $\mathrm{Prim}(G)$ and $E = E(J) \subset \widehat{L}/\widetilde{G}$.

LEMMA 3. *We assume that σ is a factor representation of D, the restriction of which to L is carried by E. Then σ is of type* I.

Since D is contained in G_π, the action of D on E is trivial. Hence we can assume that $\sigma|L$ is a multiple of π. Writing $\pi^e_1 = \pi^e|D^e$ and $\phi_1 = \phi|D^e$, we get $\sigma = (\pi^e_1 \otimes (\mu \circ \phi_1))^\sim$, where $\mu \in \mathrm{Fac}(\mathcal{E})$ is such that $\mu|\mathcal{T} \approx \overline{\mathrm{id}}$. (Here $\mathcal{T}$ stands for the canonical torus in Γ, id for the identity map of $\mathcal{T}$ onto itself.) Since, by Lemma 2, $\mu = (I)$ we can conclude that $\sigma = (I)$. Summing up we have completed proving the following: If $\sigma \in \mathrm{Fac}(D)$ is such that $\sigma|L \subset E$, then $\sigma = (I)$.

We set $F' = \{\sigma; \sigma \in \widehat{D}$ such that $\sigma|L \subset E\} \subset \widehat{D}$.

LEMMA 4. *Assume that σ is a fixed element in F'. Then $G_\sigma = D$.*

In fact,

a) We note first that D is invariant in $\widetilde{G}$. In fact, this follows from $[\widetilde{G}, \widetilde{G}] = L \subset D$. Hence we can assume that $\sigma|L \approx \pi$ (fixed at the beginning). Setting $D' = G_\sigma$, we have thus $D \subseteqq D'$ and it suffices to prove that $D = D'$.

b) We note next that D' is contained in G_π. In fact, if $a \in D'$, then $a\sigma = \sigma$, whence, since $L \subseteqq D$, $a\pi = \pi$, and thus $D' \subseteqq G_\pi$.

c) There is an element μ of $\mathrm{Irr}(\mathcal{E})$ such that, with notation used above, we get $\sigma = [\pi_1^e \otimes (\mu \circ \phi_1)]^\sim$. Next we observe that if $a \in D'$, $\gamma = \phi((a,1)) \in \Gamma$, $U = \pi^e((a,1)) \otimes I$ then we have for all $d \in D$

$$(a\sigma)(d) = [U^*(\pi_1^e(d,1)) \otimes (\gamma\mu)(\phi_1((d,1))U]^\sim.$$

In fact, given $d \in D$, we have

$$(a\sigma)(d) = \pi_1(a^{-1}da, 1) \otimes \mu(\phi_1(a^{-1}da, 1)).$$

Since $(a^{-1}da, 1) = (a,1)^{-1} \cdot (d,1)(a,1)(e,u)$ $(u \in T)$ we have that

$$(a\sigma)(d) = (U^*(\pi^e((d,1)) \otimes (\gamma\mu) \circ \phi((d,1)))U)^\sim.$$

d) We observe next that if $a\sigma = \sigma$ (a fixed in D') and $\gamma = \phi((a,1))$, then γ is in $\mathcal{E}$.

In fact, there is $\chi \in \mathrm{X}(H)$ such that $\mu|H \approx \chi$ and thus $\gamma\chi = \chi$. Given some h in H, let us set $\gamma^{-1}h\gamma = uh$ $(u \in \mathcal{T})$; it is enough to show that $u = 1$. In fact, we have

$$\chi(h) = (\gamma\chi)(h) = \chi(\gamma^{-1}h\gamma) = \overline{u}\chi(h) \Rightarrow u = 1.$$

We deduce that $\gamma \in C(H) = \mathcal{E}$.

e) Conclusion: $\phi((a,1)) \in \mathcal{E}$ implies that $(a,1)$ is in D^e and thus $a \in D$.

Summing up, we have completed proving that if a is in G and σ in F' such that $a\sigma = \sigma$ then $a \in D$.

For more details concerning the following lemma, cf. Lemma 6 in Section 3.2. We recall that E is a fixed $\widetilde{G}$ orbit in $\widehat{L}$ and we set

$$F' = \{\sigma; \sigma \in \widehat{D} \text{ such that } \sigma|L \subseteqq E\}.$$

LEMMA 5. *F', as a subspace of $\widehat{D}$, is locally compact, Hausdorff.*

a) We start by observing that F' is locally closed in $\widehat{D}$. To show this, we shall avail ourselves of Lemma 1 in 1.3):

Suppose that L is closed, invariant and of type I in G. Assume that $A \subset \widehat{L}$ is locally closed and G-invariant. Setting $\widehat{G}_A = \{\rho; \rho \in \widehat{G}$ such that $\rho|L \subseteqq A\}$, then $\widehat{G}_A$ is locally closed in $\widehat{G}$. Applying this to D, L, E in place of G, L, A, in which case we get $\widehat{G}_A = F'$, we conclude that F' is locally closed in $\widehat{D}$.

Let us write $\mathcal{D} = C^*(D)$. Then we can infer that there are $I_1 \supset I_2$, closed, 2-sided ideals in $\mathcal{D}$ such that if $M = I_1/I_2$, then $F' = \widehat{M}$.

b) We note next that M is of type I. To this end it is enough to show that if $\sigma' \in \mathrm{Fac}(I_1)$ such that $\ker(\sigma') \supsetneqq I_2$ then $\sigma' = (I)$. We recall (cf. [**D***], 2.10.4, Proposition, p. 52) that given σ' as above, there is a unitary representation σ of D, well-determined by $\sigma|I_1 = \sigma'$. In this case also: $R(\sigma') = R(\sigma)$. To prove our point, let us consider

$$\sigma = \int_X^\oplus \sigma(x) \cdot d\mu(x)$$

where (X, μ) is standard such that $\sigma(x) \in \mathrm{Irr}(D)$ almost always with respect to μ.

1) (cf. [**Dal**], p. 100) $\|\sigma(v)\| = \|\sigma(x)\|$ a.e. (μ) for all $v \in D$. Conclusion: We can assume $\ker(\sigma(x)) \supset I_2$, always.

2) Also, evidently there is a subset $N_1 \subset X$, of measure zero and such that $\sigma(x)|I_1 \not\equiv 0$ if x is in $X - N_1$. Since $F' = \{\delta; \delta \in \widehat{D}, \ker(\delta) \not\supset I_1, \ker(\delta) \supset I_2\}$, up to a set of μ-measure zero, $\sigma(x) \in F'$, whence, by Lemma 7 in Section 3.2, $\sigma|L \subset E$ and thus, by Lemma 3 above, $\sigma = (I)$.

Conclusion: We have proved that if $I_1 \supset I_2$ are closed ideals in $\mathcal{D}$ $(= C^*(D))$ such that $M = I_1/I_2$ implies $\widehat{M} = F'$, then M is of type I.

c) Let us denote by $\mathcal{F}'$ the annihilator of L in $\mathrm{X}(D)$ and put $N' = \widetilde{G} \times \mathcal{F}'$. We claim that F' itself is an N'-orbit. In fact, to this end it is enough to show that if α, β are elements of F', such that $\alpha|L = \beta|L$ ($\approx \pi_1$ say) then there is $\varphi \in \mathcal{F}'$ such that $\varphi\alpha = \beta$. In fact, let us put $\widehat{\mathcal{E}}_1 = \{\mu; \mu \in \widehat{\mathcal{E}}$ such that $\mu|\mathcal{T} \approx \overline{\mathrm{id}}\}$. Given $\eta \in \widehat{\mathcal{E}}_1$, let us set $f(\eta) = [\pi_1^e \otimes (\eta \circ \phi_1)]^{\sim}$.

Conclusion: There is $\gamma, \delta \in \widehat{\mathcal{E}}$ such that $\alpha = f(\gamma)$, $\beta = f(\delta)$.

Next we note that if $\mathcal{F}''$ is the annihilator of $\mathcal{T}$ in $\mathrm{X}(\mathcal{E})$ there is an isomorphism $\tau : \mathcal{F}' \to \mathcal{F}''$ such that $\varphi f(\eta) = f(\tau(\varphi)\eta)$ ($\eta \in \widehat{\mathcal{E}}$ fixed; $\varphi \in \mathcal{F}'$). From here, to attain our goal it suffices to show that $\mathcal{F}''$ acts transitively on $\widehat{\mathcal{E}}_1$. To this end it is enough to recall that, by Lemma 1, $\mathcal{E} = \Gamma_0 \times \Delta$ where Δ is free abelian and $\Gamma_0 = \Gamma_0' \times A$, where A is a vector group and Γ_0' is a Heisenberg group, divided by a discrete subgroup of the center.

Summing up, in c) above we have established that if $\mathcal{F}'$ is the annihilator of L in $\mathrm{X}(D)$ and we put $N' = \widetilde{G} \times \mathcal{F}'$, then F' is acted upon transitively by N'.

d) We recall that what we are to prove is that F', as a subspace of $\widehat{D}$, is locally compact and Hausdorff. In fact, we have by b) that $F' = \widehat{M}$, where $M = (I)$. Hence F' is locally quasi-compact and almost Hausdorff. We conclude from this first (cf. [**Gl**], Th. 1, (3) $\Rightarrow$ (6)) that for any $\rho \in F'$, fixed, then $nN'_\rho \mapsto n\rho$ ($n \in N'$) is a homeomorphism. Next we note that by $M = (I)$, $\widehat{M} = F'$ is standard, and so is N'/N'_ρ and thus N'_ρ is closed (cf. [**AM**], p. 16).

LEMMA 6. *There is an equivalence relation Σ' on F' such that $\rho_1 \sim \rho_2(\Sigma')$ if and only if $\overline{G\rho_1} = \overline{G\rho_2}$. Also, F'/Σ' is countably separated.*

In fact,

$$[N', N'] = [\widetilde{G}, \widetilde{G}] = L \subset D \subset N'_\rho.$$

We continue to keep fixed a primitive ideal J that is not of type I (cf. the preceding section).

Since $F' = \widehat{M}$, where M is of type I, there is a Borel cross section $\sigma : F' \to \mathrm{Irr}(D)$ (cf. [**D***], 4.6.2, p. 95).

LEMMA 7. *Suppose that $V \in \mathrm{Facn}(G)$ is given such that $\ker(V) = J$. We claim that there is on F' a measure, ν, say, quasi-invariant and ergodic with respect to the action of G on F', such that*

$$V|D = \int_{F'}^{\oplus} \sigma'(\lambda) \cdot d\nu(\lambda), \qquad \sigma(\lambda) \approx \sigma'(\lambda).$$

a) Below we shall imitate the reasoning of a) in Lemma 1, Section 3.2. We start by recalling that there is a standard measure μ on $\widehat{D}$ and a μ-measurable section $V : \widehat{D} \to \mathrm{Fac}(D)$ such that

$$V|D = \int_{\widehat{D}}^{\oplus} V(\zeta) \cdot d\mu(\zeta)$$

(cf. [**D***], 4.8.2, Théorème, p. 147). We claim that there is a μ-measurable subset $A \subset \widehat{D}$, the complement of which is of a μ-measure zero, such that if $\zeta \in A$, then $V(\zeta)|L$ is carried by E. In fact, we have

$$V|L = \int_{\widehat{D}}^{\oplus} (V(\zeta)|L) \cdot d\mu(\zeta).$$

Since $\ker(V) = J$, we have $V|L \subset E$ $(= E(J)) \in \widehat{L}/\widetilde{G}$, and thus, by Lemma 7 in Section 3.2, $V(\zeta)|L \subseteqq E$ almost everywhere with respect to μ. Evidently, we can assume that A is G-invariant.

b) We write $g : \widehat{D} \to \widehat{D}_I$ for the canonical inclusion. We recall (cf. [**D***], 7.3.6, Théorème, p. 139) that $\widehat{D}_I$ is a Borel subset of $\widehat{D}$, and g is a Borel isomorphism with its image. We claim that A is contained in $g(F')$. In fact, since $V(\zeta)|L \subseteqq E$ for any $\zeta \in A$, it is enough to recall that if $\sigma \in \mathrm{Fac}(D)$ such that $\sigma|L \subseteqq E$, then σ is of type I.

c) We define a Borel measure ν on F' such that the image, through g, of its part in $g^{-1}(A)$ is equal to $\mu|A$, and the complement be zero. In this manner we obtain

$$V|D = \int_{F'}^{\oplus} \sigma'(\lambda) \cdot d\nu(\lambda), \qquad \sigma'(\lambda) \approx \sigma(\lambda).$$

This decomposition is central and, by $V \in \mathrm{Fac}(G)$, it is G-quasi-invariant and ergodic.

Below unexplained notation will be as above.

LEMMA 8. *Let V be a normal factor representation of G with the kernel J. We claim that there is a unitary representation W of D such that V is quasi-equivalent to* $\mathrm{ind}_{D\uparrow G} W$.

a) Since in Lemma 7 ν is, in particular, ergodic with respect to G, we can replace F' by some $\mathcal{A} \in F'/\Sigma'$ (cf. Lemma 6), obtaining thus

$$V|D = \int_{\mathcal{A}}^{\oplus} \sigma'(\lambda) \cdot d\nu(\lambda), \qquad \sigma'(\lambda) \approx \sigma(\lambda).$$

b) Let N' be as in the proof of Lemma 5, and let x be fixed in $\mathcal{A}$. Putting $H = \overline{GN'_x}$, we have clearly $\mathcal{A} = Hx$. We note, incidentally, that we have $H_x \cap G = G_x = D$. Before proceeding we introduce the following notation. If A is some subgroup of B, we write $\mathrm{X}(B/A)$ for the annihilator of A in $\mathrm{X}(B)$.

We put $C = \mathrm{X}(G/D)$ and C' for the totality of all those elements ω in C, for which there is some ω' in $\mathrm{X}(H/H_x)$ such that ω is the restriction to G of ω'.

We note that C' is dense in C. In fact, if not, then there is $g \in G - H_x$ such that $\omega(g) \equiv 1$ $(\omega \in C')$. But, evidently, given g in $G - H_x$, there is an element ω in $\mathrm{X}(H)$ such that $\omega|H_x \equiv 1$ and $\omega(g) \neq 1$.

c) Suppose that ω is some element in C'. We claim that there is a continuous map $\psi : \mathcal{A} \to \mathbb{C}$ such that $|\psi| \equiv 1$ and that $\psi(ay) = \omega(a)\psi(y)$ $(a \in G, y \in \mathcal{A})$. In fact, suppose that φ in $\mathrm{X}(H/H_x)$ restricts to ω in $\mathcal{A}$. Assume furthermore that for y in $\mathcal{A}$, there are elements a, b in H such that $ax = y = bx$. We observe that $\varphi(a) = \varphi(b)$. In fact, this is so, since $b = ad$ $(d \in H_x)$. In this manner we can define a compiex-valued function ψ on $\mathcal{A}$ such that $\psi(y) = \omega(a)$ if $y = ax$. We have, evidently, that if y is in $\mathcal{A}$ and g in G, then $\psi(gy) = \omega(g)\psi(y)$ and $|\psi| \equiv 1$.

d) We recall that

$$V|D = \int_{\mathcal{A}}^{\oplus} \sigma'(\lambda)\, d\nu(\lambda).$$

We define $V(\psi) = \int_{\mathcal{A}}^{\oplus} \psi(x) I_x \, d\nu(x)$. Then $V(\psi)$ is a unitary operator in the center of $R(V|D)$. One sees easily that if $g \in G$, then $V(g)V(\psi)V(g^{-1}) = \overline{\omega(g)}V(\psi)$.

e) Let us put $M = R(V)$. Since V is a factor representation, which is normal, there is a faithful, normal, semifinite trace on M. We claim that if $\omega \in C$ is fixed, then we have for all a in $\mathfrak{G}^+ : \phi(V(\omega a)) = \phi(V(a))\ (\leqq +\infty)$. In fact, assume first that ω is in C'. We have, by d), that $V(g)V(\psi)V(g^{-1}) = \overline{\omega(g)}V(\psi)$, whence $(V(\psi))^* V(g) V(\psi) = \overline{\omega(g)} V(g)$. If f is in $L^1(G)$, then $(V(\psi))^* L(f) V(\psi) = V(\overline{\omega} f)$ implying for all b in $\mathfrak{G} : (V(\psi))^* V(b) V(\psi) = V(\overline{\omega} b)$. Summing up, what we have said so far implies that for all ω in C' and a in $\mathfrak{G}$, we have $(V(\psi))^* \cdot V(a) \cdot V(\psi) = V(\overline{\omega} a)$. Since $V(\psi)$ is unitary in $R(V)$ this entails that if a is in $\mathfrak{G}^+$, then $\phi(V(a)) = \phi(V(\omega a))\ (\leqq +\infty)$. By the normalcy of ϕ, and since C' is dense in C (cf. b), end) the same holds true for all $\omega \in C$ and $a \in \mathfrak{G}^+$.

f) 1) For the following cf. [**D vN**], Ch. I, §6, pp. 85–89. Let $\mathfrak{m}$ be the collection of all operators in M for which $\phi(A^* \cdot A) < +\infty$. Then $\mathfrak{m}$ can be endowed with the structure of a unitary algebra such that $X, Y \in \mathfrak{m}$ implies: $(X, Y) = \phi(X \cdot Y^*)$. Let $\mathfrak{h}$ be the completion of $\mathfrak{m}$. Let us put

$$\mathfrak{n} = \{a; a \in \mathfrak{G} \text{ s.t. } \phi(V(a^* \cdot a)) < +\infty\}.$$

We observe that if $\mathfrak{h}' = \{V(a); a \in \mathfrak{m}\} \subset \mathfrak{h}$ then $\mathfrak{h}'$ is dense in $\mathfrak{h}$. Let us write P for the projection onto $\overline{\mathfrak{h}}'$. We have $V(\mathfrak{G})\mathfrak{h}' \subseteqq \mathfrak{h}', \mathfrak{h}'V(\mathfrak{G}) \subseteqq \mathfrak{h}'$, and thus, since M is a factor and ϕ is normal we obtain $P = I$ and hence $\overline{\mathfrak{h}}' = \mathfrak{h}$.

2) We note that, evidently, there is a unitary representation V' of G on $\mathfrak{h}$, well-determined by $V'(g)(T) = V(g)T\ (T \in \mathfrak{h}')$.

3) By e), there is a unitary representation Z of C on $\mathfrak{h}$, such that for all ψ in C and a in $\mathfrak{n}$ we have: $Z(\psi)V(a) = V(\psi a)$.

4) We have for all g in G and ψ in C:

$$V'(g)Z(\psi)V'(g^{-1}) = \overline{\psi(g)}Z(\psi).$$

5) V' is quasi-equivalent to V.

g) We recall that our objective is to establish the existence of a unitary representation W of D such that V is quasi-equivalent to the representation induced by W in G. Putting $\overline{G} = G/D$ we note that $\overline{G}$ is the dual of $C = \mathrm{X}(G/D)$. Hence there is a spectral measure $\mathcal{P}$ on $\overline{G}$ such that

$$Z(\psi) = \int_{\overline{G}} \psi(h) \cdot d\mathcal{P}(h) \qquad (\psi \in C).$$

Given g in G, we write $\overline{g}$ for its image in $\overline{G}$. Let E be some Borel set in $\overline{G}$. We have evidently $V'(g)\mathcal{P}(E)V'(g^{-1}) = \mathcal{P}(E\overline{g})$. Thus $(V', \mathcal{P})$ is a system of imprimitivity based on $\overline{G}$.

Thus, by [**Mackact**], Theorem 6.6, p. 291, there is a unitary representation W of D such that $V' = \mathrm{ind}_{D \uparrow G} W$. In this manner, to complete our demonstration it is enough to note that V' is quasi-equivalent to V.

Before proceeding with the discussion of our main topic, we have to recall a few things concerning the representation theory of a separable C^*-algebra A.

1) Given two representations T and T' we say that they are disjoint if and only if no subrepresentation of T is unitarily equivalent to some subrepresentation of T'.

2) Given a *-representation T of A, we say that it has property (1) if $(T(A))'$ is abelian. On the other hand, it will be said to admit the property (2) in the case that, in a direct sum decomposition $T = T_1 \oplus T_2$, T_1 and T_2 are disjoint.

All this being so, (1) and (2) can be shown to be equivalent, and a *-representation will be said to be without multiplicity (or multiplicity free) if it admits either of these properties. From our present position two facts will play an important role:

1) If T_1 and T_2 are quasi-equivalent and both are without multiplicity then they are equivalent (cf. [**D***], 6.4.6, p. 108).

2) A representation T is without multiplicity if and only if its central decomposition is carried by $\widehat{A} = \widehat{A}_I \subset \widehat{A}$, and there is of the form

$$T = \int_{\widehat{A}}^{\oplus} T(\zeta) \cdot d\mu(\zeta)$$

where $T(\zeta)$ is irreducible of class ζ (cf. [**D***], 8.4.8, p. 162).

LEMMA 9. *Let T be a normal factor representation of G such that $T = \operatorname{ind}_{D\uparrow G} W$ (cf. Lemma 8). Then*

(i) *W can be chosen to be G-invariant and multiplicity free.*

(ii) *If W_1 and W_2 are as in* (i), *and $\operatorname{ind}_{D\uparrow G} W_1$ is quasi-equivalent to $\operatorname{ind}_{D\uparrow G} W_2$, then W_1 and W_2 are equivalent.*

In fact,

(i) Let us put $W_1 = \int_g^{\oplus} gW \cdot dg$, where dg is a Haar measure on G. Then $\operatorname{ind}_{D\uparrow G} W_1 = (+\infty)(\operatorname{ind}_{D\uparrow G} W) \approx T$, and thus, replacing W by W_1, we can assume that W is G-invariant.

Below we shall use freely the notation and results of Lemmas 5–8.

Then we have, by Lemma 7,

$$(+\infty)W = T|D = \int_{\mathcal{A}}^{\oplus} \sigma'(x) \cdot d\nu(x)$$

where ν is G-quasi-invariant and ergodic. It is important to recall (cf. Lemma 5) that $\mathcal{A} \subset F' = \widehat{M}$, and M is of type I; $\sigma(x)$ is irreducible of class x.

We can write

$$W = \int_{\mathcal{A}}^{\oplus} n(x)\sigma(x)\, d\mu(x)$$

where $n : \mathcal{A} \to \mathbb{Z}$ is a positive integral-valued measurable function. Hence we have $\nu = \mu$ and $n(x) = N$, where N is some constant. We write

$$U = \int_{\mathcal{A}}^{\oplus} \sigma(x) \cdot d\mu(x).$$

Then $W = N \cdot U$, U is multiplicity free and $\operatorname{ind}_{D\uparrow G} U \approx \operatorname{ind}_{D\uparrow G} W$.

(ii) Here we have to prove that if W_1 and W_2 are G-invariant and multiplicity free such that

$$\operatorname{ind}_{D\uparrow G} W_1 \approx \operatorname{ind}_{D\uparrow G} W_2$$

then W_1 and W_2 are equivalent. In fact,

$$(+\infty)W_1 = \Big(\operatorname{ind}_{D\uparrow G} W_1\Big)|D \approx \Big(\operatorname{ind}_{D\uparrow G} W_2\Big)|D = (+\infty)W_2.$$

Here we used the fact that if T_1 and T_2 are quasi-equivalent, then so are their restrictions on some subgroup. In fact, our assumption implies that there exist positive integers n, m such that nT_1 and mT_2 are equivalent.

Summing up, $(+\infty)W_1$ is quasi-equivalent to $(+\infty)W_2$ and thus they are also equivalent (cf. [**D***], 5.4.6, p. 108).

We write $\widetilde{N} = \widetilde{G} \times \mathrm{X}(G)$. If $\chi \in \mathrm{X}(G)$, then let $\tau(\chi)$ be $\chi|D$. We recall that $N' = \widetilde{G} \times \mathcal{F}'$, where $\mathcal{F}' = \mathrm{X}(D/L)$. Writing $j(a,\chi) = (a, \tau(\chi)) \in N'$, the map $j : \widetilde{N} \to N'$ is a surjective morphism.

From this we conclude that $\widetilde{N}$ acts on $\widehat{D}$ and on F'. We recall (cf. Lemma 5) that F' itself is an $\widetilde{N}$-orbit.

Let us fix an element σ in F'. Then we have $[\widetilde{N}, \widetilde{N}] = [\widetilde{G}, \widetilde{G}] = L \subset D \subset \widetilde{N}_\sigma$ (cf. Lemma 6); hence $\widetilde{N}_\sigma$ ($\sigma \in F'$) is constant. If $\Lambda \equiv \widetilde{N}_\sigma$ ($\sigma \in F'$) then $[\widetilde{N}, \widetilde{N}] \subseteqq \Lambda$.

Conclusion: $A = \widetilde{N}/\Lambda$ is a connected abelian Lie group, and $a\Lambda \mapsto a\sigma$ is an A-equivariant homeomorphism. If $\mathcal{A} \in F'/\Sigma'$ (cf. Lemma 6), let us put $\mathcal{E} = \{\alpha; \alpha \in \widetilde{N}$ such that $\alpha\mathcal{A} \subset \mathcal{A}\}$. Then $\mathcal{E}$ is the inverse image of the closure of the image of G in A. Let Γ be the image of $\mathcal{E}$ in $\mathcal{A}$; then $\mathcal{A}$ as above is just a translate of Γ.

Before proceeding with the substance of our discussion, we make the following remarks.

1) Let us write $\mathfrak{N}$ for the totality of the equivalence classes of the W' as above. We define $\eta : \widehat{G}_{\text{norm}} \to \mathfrak{N}$ as follows. If l is in $\widehat{G}_{\text{norm}}$, $\mathrm{cl}(T) = l$, $T \approx \mathrm{ind}_{D\uparrow G} W$, where W is G-invariant and multiplicity free, then $\eta(l) = \mathrm{cl}(W)$.

2) η is injective.

In fact, assume that l_1, l_2 are in $\widehat{G}_{\text{norm}}$ such that $\eta(l_1) = \eta(l_2)$. Assume that $\mathrm{cl}(T_j) = l_j$ $(j = 1, 2)$ and $T_j \approx \mathrm{ind}_{D\uparrow G} W_j$. The assumption implies that $W_1 \sim W_2$ and thus evidently $T_1 \approx T_2$ and $l_1 = l_2$.

3) We observe that if $T_1 \approx T_2$ and $\alpha \in \mathcal{E}$, then $\alpha T_1 \approx \alpha T_2$. In fact, if $T_j = \mathrm{ind}_{D\uparrow G} W_j$, then $W_1 \sim W_2$ and thus $\alpha W_1 \sim \alpha W_2$, and hence

$$\begin{aligned} \alpha T_1 &\approx \alpha\Big(\mathrm{ind}_{D\uparrow G} W_1\Big) = \mathrm{ind}_{D\uparrow G} \alpha W_1 \sim \mathrm{ind}_{D\uparrow G} \alpha W_2 \\ &= \alpha\Big(\mathrm{ind}_{D\uparrow G} W_2\Big) \approx \alpha T_2. \end{aligned}$$

We have thus proved that if $T_1 \approx T_2$, $\alpha \in \mathcal{E}$ then $\alpha T_1 \approx \alpha T_2$. In particular, given $l \in \widehat{G}_{\text{norm}}$, and $\alpha \in \mathcal{E}$ we can define αl.

4) We note next that if $l \in \widehat{G}_{\text{norm}}$ and $\alpha \in \mathcal{E}$ then $\alpha\eta(l) = \eta(\alpha l)$. In fact, if $T \approx \mathrm{ind}_{D\uparrow G} W, \alpha \in \mathcal{E}$, then $\alpha T \approx \mathrm{ind}_{D\uparrow G}(\alpha W)$ whence

$$\eta(\alpha l) = \mathrm{cl}(\alpha W) = \alpha\eta(l).$$

In this manner we have proved that if $l \in \widehat{G}_{\text{norm}}$ and $\alpha \in \mathcal{E}$ then $\alpha\eta(l) = \eta(\alpha l)$.

5) Assume that $l \in \widehat{G}_{\text{norm}}$, $\mu = \eta(l)$.

We claim that $\mathcal{E}_l = \mathcal{E}_\mu$. In fact, $\alpha \in \mathcal{E}_l$ implies $\alpha l = l$ whence $\alpha\eta(l) = \eta(\alpha l) = \eta(l)$, and thus $\alpha \in \mathcal{E}_\mu$. If, on the other hand, $\alpha \in \mathcal{E}_\mu$, then $\alpha\eta(l) = \eta(l)$ and hence, by 4), $\eta(\alpha l) = \eta(l)$, in which manner, since η is injective, $\alpha l = l$ and $l \in \mathcal{E}_l$. We conclude from all this that $\mathcal{E}_l = \mathcal{E}_\mu$.

Assume that $J \in \mathrm{Prim}(G)$ is such that $J \neq (I)$. By Lemmas 7 and 9, there is an F' as in Lemma 7, a G-quasi-orbit, living on some $\mathcal{A}$ in F'/Σ', well determined by J, such that if $\mathrm{cl}(W) \in \mathfrak{N}$, then with notation as loc. cit.

$$W = \int_{\mathcal{A}}^{\oplus} \sigma(\lambda) \cdot d\nu(\lambda).$$

(A quasi-orbit is defined as in [**Mackact**], p. 295: an ergodic invariant measure class.)

We write $\mathfrak{M}$ for the collection of all measure class on $\mathcal{A}$, arising from members of $\mathfrak{N}$ in the said manner. The map $\varepsilon : \mathfrak{N} \to \mathfrak{M}$ is an $\mathcal{E}$-equivariant injection. Putting $\omega = \varepsilon \circ \eta$, then ω, too, will have these properties.

Objective below. Given $l_j \in \widehat{G}_{\mathrm{norm}}$ such that $\ker(l_1) = \ker(l_2)$, then $l_1 = l_2$. We shall do this by proving that, given $l \in \widehat{G}_{\mathrm{norm}}$, $\omega(l)$ contains a well-determined element.

LEMMA 10. *If $l \in \widehat{G}_{\mathrm{norm}}$, then its stabilizer in $\widetilde{N}$ is closed.*

In fact, we start by recalling (cf. [**AM**], Proposition 3.7, p. 16) that if G is separable and locally compact, E a countably separated Borel space, acted upon by G in such a manner that for each $x \in E$ the map $a \mapsto ax$ from G into E is Borel, then G_x is closed. Next we remark (cf. [**Hal**]) that $\widehat{G}_{\mathrm{norm}}$ is standard. In this manner it is enough to note that for any V fixed in $\mathrm{Facn}(G)$, the map $\alpha \mapsto \alpha V$ ($\alpha \in \widetilde{N}$) is Borel.

LEMMA 11. *Assume that μ is some element in $\mathfrak{M}$. Then μ is the translate of the Haar measure on Γ.*

In fact, by Lemma 10, $\mathcal{E}_\mu$ is closed. Also $\Lambda \subseteqq \mathcal{E}_\mu$, $G \subseteqq \mathcal{E}_\mu$, whence

$$\mathcal{E} \subseteqq \overline{\Lambda G} \subseteqq \mathcal{E}_\mu \subseteqq \mathcal{E}$$

implying $\mathcal{E}_\mu = \mathcal{E}$. Hence it is enough to recall (cf. [**Macktr**], Lemma 7.3, p. 145) that if Γ is a locally compact group and μ is quasi-invariant with respect to Γ, then μ is equivalent to the Haar measure on Γ.

PROPOSITION. *Assume that $l_1, l_2 \in \widehat{G}_{\mathrm{norm}}$ are such that $\ker(l_1) = \ker(l_2)$. Then $l_1 = l_2$.*

In fact, by Lemma 11, $\omega(l_1) = \omega(l_2)$ whence, since ω is injective, $l_1 = l_2$.

3.8. Proof of principal result

Below our objective will be to prove the principal result of this chapter:

THEOREM. *Let G be a connected Lie group, J a primitive ideal of the group C^*-algebra. Then there is an up to quasi-equivalence, well-determined, normal factor representation, admitting J as its kernel.*

We start by observing that by the proposition in Section 3.3 and that of Section 3.7 we have already established our claim for the special case when our group is simply connected. Let G' be the universal covering of G, $\phi : G' \to G$ the canonical morphism, and $\Gamma = \ker(\phi)$. In general, notions referring to G' will be distinguished with a prime. Let V_j ($j = 1, 2$) be normal representations of G such that $\ker(V_1) = \ker(V_2)$. We set $V_j' = V_j \circ \phi$. Our assumption implies that $\ker(V_1') = \ker(V_2')$ and conversely. We conclude from this that given a primitive ideal J in $\mathrm{Prim}(G)$, there is a unique $J' \in \mathrm{Prim}(G')$ such that if T is a representation of G with a kernel equal to J, then $\ker(T') = J'$ and conversely, if U satisfies $\ker(U) = J'$, then $U = V'$ for some V on G.

All this being so, it is enough to show that if $\ker(V_1') = \ker(V_2')$, then $V_j' \in \mathrm{Facn}(G')$.

Let $V = V_1'$, say; to prove our point it is enough to show that $V(\mathfrak{G}) = V'(\mathfrak{G}')$.

Let $d\gamma$ be the Haar measure on Γ. Given $f' \in \mathcal{L}(G')$ we write $\tau(f)(x) \equiv \sum_{\gamma\in\Gamma} f(x'\gamma)$. Then $\tau(\mathcal{L}'(G')) = \tau(\mathcal{L}(G))$ and thus $V(\tau(\mathcal{L}(G)) = V'(\mathcal{L}(G'))$ whence finally $V(\mathfrak{G}) = V'(\mathfrak{G}')$.

3.9. The theorem of Poguntke

A more recent approach, involving ideas of substantial depth, to the theorem of Section 3.8 was given by D. Poguntke. We confine ourselves here to quoting his relevant principal result (cf. [**Pog**], Theorem 2, p. 161).

Let G be a connected and simply connected Lie group. Then any primitive subquotient of $C^*(G)$ contains a unique simple, closed ideal, which is isomorphic either to a finite-dimensional full matrix algebra, or to $C(H) \otimes C^*(F, \alpha)$ where $C(H)$ is the algebra of all compact operators of a separable Hilbert space H, F a free abelian group and α a nondegenerate cocycle on $F \times F$.

The theorem of Section 3.8 can be obtained as a fairly simple corollary of the above statement.

CHAPTER IV

Generalized Kirillov Theory

The objective of this chapter is the description, given an arbitrary connected and simply connected solvable Lie group, of its primitive ideal space by aid of the Lie algebra and the coadjoint action on it. By doing so we extend previous work by Kirillov, Auslander and Kostant, to whom we are much indebted.

Our task requires the introduction of several new concepts. Let G be our group with the Lie algebra $\mathfrak{g}$. We consider the dual $\mathfrak{g}^*$ of the underlying space of $\mathfrak{g}$ as a G-module with respect to the coadjoint representation. Given an element g of $\mathfrak{g}^*$ we form its stabilizer G_g in G. Since G is assumed, in particular, to be solvable, the connected component of the identity $(G_g)_0$ is itself simply connected. Therefore, there is a character χ_g of $(G_g)_0$ well-determined by the condition that its differential be equal to $-2\pi i(g|\mathfrak{g}_g)$. During our subsequent considerations a significant role will be played by the question as to how χ_g extends to a larger subgroup of G_g. To clarify the situation, we introduce the concept of the reduced stabilizer of g (cf. 4.2 below). This is a subgroup of G_g, containing $(G_g)_0$, to which χ_g admits an extension; we denote it by $\overline{G}_g$. Writing $\overset{A}{\mathcal{G}}_g$ for the collection of all extensions of χ_g to $\overline{G}_g$, we form the disjoint union

$$\mathcal{R} = U_{g\in\mathfrak{g}^*}\overset{A}{\mathcal{G}}_g.$$

A typical element of $\mathcal{R}$ is a pair, composed of an element g of $\mathfrak{g}^*$ and a character, extending χ_g to $\overline{G}_g$. $\mathcal{R}$ carries naturally the structure of a G-space; the projection of this action onto $\mathfrak{g}^*$ is just the coadjoint representation. This being so, given p in $\mathcal{R}$, we shall introduce a canonical unitary equivalence class $F(p)$ of semifinite factor representations. Members of $F(p)$ are obtained, through holomorphic induction (cf. 4.2), from one-dimensional characters. If G is nilpotent, G_g is always connected; thus $\overset{A}{\mathcal{G}}_g$ is a one-element set and $F(p)$ is composed of multiples of the irreducible unitary representation corresponding, in the sense of Kirillov, to the coadjoint orbit of g. For p, q in $\mathcal{R}$, $F(p)$ and $F(q)$ are unitarily equivalent if and only if p and q lie on the same G-orbit (cf. 4.7 below). Given p in $\mathcal{R}$, we write $J(p)$ for the primitive ideal given by the kernels of members of $F(p)$. This being so the map $J : \mathcal{R} \to \mathrm{Prim}(G)$ is surjective (cf. 4.11). For a given primitive ideal, its preimage in $\mathcal{R}$ underlies a differentiable manifold O. In more detail, this is a foliation, where the leaves are isomorphic to coadjoint orbits of G. We shall call O a generalized orbit. In the nilpotent case, this is just a coadjoint orbit. Given O, one can show the existence of a measurable section $p \mapsto T(p) \in F(p)$.

Let v be an absolutely continuous measure on O, and form the representation

$$T = \int_O^{\oplus} T(p) \cdot dv(p).$$

We show that this is a factor representation, the kernel of which is equal to the common values of $\ker(T(p))$ $(p \in O)$. The principal result of the chapter (4.12, Theorem) states that T is a normal representation. Thus we obtain all elements of $\widehat{G}_{\text{norm}}$ by letting O describe the totality of all generalized orbits. Again, if G is nilpotent, as we said earlier, O is a coadjoint orbit and thus T is a multiple of an irreducible representation. We give a necessary and sufficient condition, in terms of the geometry of O, that the corresponding primitive ideal be of type one (cf. 3.6 and 4.13). This permits a new derivation of the theorem of Auslander and Kostant saying that G is of type one if and only if any O (as above) is G-transitive and for any g, in the projection of O into $\mathfrak{g}^*$, the reduced stabilizer coincides with the stabilizer itself.

4.1. Preliminary facts

In this section we collect a few facts necessary for the explanation of our version of real and holomorphic induction. They are mostly well known.

Let G be a separable locally compact group and A a closed subgroup. Let dg and da be right-invariant Haar measures such that $d(g_0 \cdot g) = \Delta_G(g_0) \cdot dg$ and $d(a_0 \cdot a) = \Delta_A(a_0) \cdot da$. We denote by $\mathcal{L}(G)$ the set of all continuous functions with compact support on G. Given $f \in \mathcal{L}(G)$ we put

$$F(x) = \int_A f(ax) \cdot da.$$

Then $F \in \mathcal{L}(G/A)$ and the map $f \mapsto F$ is surjective. Let $d\nu$ be a positive Borel measure on G satisfying $d\nu(ax) = \Delta_A(a) \cdot d\nu(x)(a \in A)$. Then there is a Borel measure μ on G/A, well-determined by the following: If $f \in \mathcal{L}(G)$, then

$$\int_{G/A} F(p) \cdot d\mu(p) = \int_G f(x) \cdot d\nu(x).$$

We write $\mu = \nu/A$. Let us define for $a \in A : \eta(a) \equiv \Delta_A(a)/\Delta_G(a)$. If $l(x) \geqq 0$ is locally integrable such that $l(ax) \equiv \eta(a)l(x)$ $(a \in A, x \in G)$, then putting $d\nu(x) = l(x) \cdot dx$, we have $d\nu(ax) = \Delta_A(a) \cdot d\nu(x)$.

Let φ be a character of A. We proceed to define the representation $U = \operatorname{ind}_{A\uparrow G} \varphi$, given rise to by φ of G, in the following manner. We consider the collection $\mathcal{E}$ of all complex-valued measurable functions on G satisfying for all $x \in G, a \in A$:

$$f(ax) = (\eta(a))^{1/2} \cdot \varphi(a) f(x)$$

and for which $|f(x)|^2$ is locally integrable with respect to dx. We can define $d\mu_f = |f|^2 \cdot dx/\Delta$. Consider the collection $\mathcal{F}$ of all those functions f in $\mathcal{E}$ for which $\mu_f(G/A)$ is finite. Let us write $\mathcal{F}'$ for the subset of all elements in $\mathcal{F}$ satisfying $\mu_f(G/A) = 0$. We define on $\mathcal{F}/\mathcal{F}'$ a Hilbert space structure such that if f is representing a class of $\mathcal{F}/\mathcal{F}'$, then

$$\|f\|^2 = \mu_f(G/A).$$

For $f \in \mathcal{F}$ and $g \in G$ fixed, we put $f_g(x) \equiv f(xg)$. We have evidently $f_g \in \mathcal{E}$ and $\mu_f(G/A) = \mu_{f_g}(G/A)$. Therefore, if $f \in \mathcal{F}$, then also $f_g \in \mathcal{F}$ and $\|f_g\| = \|f\|$, and $f \mapsto f_g$ $(f \in \mathcal{F})$ defines a continuous unitary representation U of G on $\mathcal{F}/\mathcal{F}'$. Let k be a continuous morphism of G into the set of all positive numbers such that $k|A \equiv \eta$.

LEMMA 1. *With the above notation, there is a positive measure* dv *on* G/A *such that* $dv(pg) = k(g) \cdot dv(p)$.

In fact,

(i) We observe that putting $d\nu(x) = k(x) \cdot dx$, we have, for any $a \in G$, $d\nu(ax) = \Delta_A(a) \cdot d\nu(x)$. Therefore we can form $k \cdot dx/A = dv$ (say).

(ii) Suppose that g is some element of G. We claim that $dv(pg) = k(g) \cdot dv(p)$. In fact, let f be some function from $\mathcal{L}(G)$. We have then

$$\int_{G/A} F(p) \cdot dv(pg) = \int_{G/A} F(pg^{-1}) \cdot dv(p) = \int_G f(xg^{-1}) \cdot k(x)\, dx.$$

By $x \mapsto xg$, this yields

$$k(g) \int_G f(x) \cdot k(x) \cdot dx = k(g) \int_{G/A} F(p) \cdot dv(p),$$

so that

$$\int_{G/A} F(p) \cdot dv(pg) = k(g) \cdot \int_{G/A} F(p) \cdot dv(p).$$

Hence, if $dv = k \cdot dx/A$, then we have

$$dv(pg) = k(g) \cdot dv(p).$$

LEMMA 2. *Let* f *be some element of* $\mathcal{F}$. *Then*

1) *the function* $x \mapsto |f(x)|^2/k(x)$ *is invariant under translation on the left by elements of* A.

2) *We have*

$$d\mu_f(x) = (|f(x)|^2/k(x)) \cdot dv(p).$$

In fact, the first statement being trivial, we turn to the second. If h is some element in $\mathcal{L}(G)$, then we have

$$\int_{G/A} H(p) \cdot d\mu_f(p) = \int_G h(x)|f(x)|^2 \cdot dx.$$

4.2. Construction of holomorphic representations

In this section our objective will be an outline of the construction of holomorphic representations; for more details cf. [**Lp 2**] or [**Dh**]. Here we shall mostly follow the second.

I. *Polarizations*

Let $\mathfrak{g}$ be a real Lie algebra of finite dimension. For a fixed g in $\mathfrak{g}^*$, we shall call the subalgebra $\mathfrak{h}$ of $\mathfrak{g}_{\mathbb{C}}$ a polarization with respect to g, or $\mathfrak{h} = \mathrm{pol}(g)$ if:

1) $\mathfrak{h}$ is a complex subalgebra of $\mathfrak{g}_{\mathbb{C}}$ such that $\mathfrak{h} + \overline{\mathfrak{h}}$, too, is a subalgebra.

2) Writing $B(x, y) = ([x, y], g)$ for x and y in $\mathfrak{g}_{\mathbb{C}}$, $\mathfrak{h}$ is self-orthogonal with respect to B.

3) If d is $\mathfrak{h} \cap \mathfrak{g}$, and $x, y \in \mathfrak{g}$ are such that $x + iy \in \mathfrak{h}$ then $B(x, y) \geqq 0$, and $B(x, y) = 0$ if and only if x, y are in d.

4) ρ is of a maximal dimension among the subalgebras satisfying 1–3.

Below, given a real Lie algebra $\mathfrak{g}$ of a finite dimension, G will stand for the corresponding connected and simply connected Lie group. We shall consider $\mathfrak{g}$ as an $\mathrm{Aut}(\mathfrak{g})$-module and $\mathfrak{g}^*$ (dual of the underlying space of $\mathfrak{g}$) as an $\mathrm{Aut}(\mathfrak{g})$-module with respect to the corresponding contragredient action.

II. *Existence of polarizations* (cf. [**AK**], Lemma II.3.1)

Let $\mathfrak{g}$ be a real nilpotent Lie algebra and g be a fixed element of $\mathfrak{g}^*$. Let K be a subgroup of $\mathrm{Aut}(\mathfrak{g})$ such that $[K, K] \subset \mathrm{Ad}(G)$ and $kg \equiv g$ $(k \in K)$. Then there is a polarization invariant under the action of K.

From now on we shall assume that $\mathfrak{g}$ is solvable.

III. *Intermezzo*: *The reduced stabilizer*

Assume that g is a fixed element of $\mathfrak{g}^*$; we write G_g for the stabilizer of g in G. We note that $G_g/(G_g)_0$ is free abelian. In fact, putting $L = [G, G]$, let us write $\mathcal{D}$ for the corresponding subalgebra of $\mathfrak{g}$. We have $\mathcal{D} = [\mathfrak{g}, \mathfrak{g}]$. Since L acts on $\mathfrak{g}^*$ with a nilpotent differential, Lg is closed in $\mathfrak{g}^*$ and thus so is LG_g in G. Hence LG_g/L is a vector group times a free abelian group. Since $L \cap G_g$ is connected, it is equal to $L \cap (G_g)_0$. We conclude from here that $G_g/(G_g)_0$, being equal to the quotient of LG_g/L by its connected component, is free abelian. We put $\mathfrak{g}_g = \mathrm{Lie}(G_g)$. The following remark is pivotal for the whole chapter. Since any connected subgroup of a connected and simply connected solvable Lie group is itself closed and simply connected, there is $\chi_g \in \mathrm{X}((G_g)_0)$ such that

$$\chi_g(\exp(l)) = \exp(-2\pi i(l, g)) \qquad (l \in \mathfrak{g}_g).$$

The reduced stabilizer $\overline{G}_g$ of g is defined as

$$\overline{G}_g = \{a; a \in G_g \text{ such that } \chi_g([a, b]) \equiv 1 \text{ for all } b \text{ in } G_g\}.$$

Evidently, $\overline{G}_g$ contains $(G_g)_0$.

As we shall see below, $g|\mathcal{D}$ determines an irreducible unitary representation π of L; the corresponding group K_π (cf. 3.1) will turn out to be equal to $\overline{G}_g \cdot L$.

To build up the normal representations of G, we shall be interested in certain unitary representations of $L\overline{G}_g$, to the description of which we now turn.

We put

$$\overset{A}{\mathcal{G}}_g = \{\chi; \chi \in \mathrm{X}(\overline{G}_g) \text{ such that } \chi|(G_g)_0 \equiv \chi_g\}.$$

By what we saw above, $\overset{A}{\mathcal{G}}_g$ is acted upon simply transitively by a torus. Often $\overset{A}{\mathcal{G}}_g$ will stand for the set of the pairs $(g, \chi), = p$ (say); $\mathrm{pol}(p)$ will mean a polarization in $\mathcal{D}_\mathbb{C}$ $(\mathcal{D} = [\mathfrak{g}, \mathfrak{g}])$ with respect to $f = g|\mathcal{D}$.

IV. *Holomorphically induced representations*

Let $g \in \mathfrak{g}^*$ be given; we put $f = g|\mathcal{D}$ and assume $f \neq 0$. Let $\mathfrak{h} = \mathrm{pol}(f) \subset \mathcal{D}_\mathbb{C}$ be invariant with respect to the action of $\overline{G}_g$ (cf. II). We write D for the connected subgroup, determined by $d = \mathfrak{h} \cap \mathcal{D}$, of L. Evidently $\overline{G}_g$ normalizes D, and $\overline{G}_g D$ is a closed subgroup of G, $= A$ say.

There is a character χ_f of D such that

$$\chi_f(\exp(l)) \equiv \exp(-2\pi i(l, f)) \qquad (l \in d).$$

Assume that χ is in $\overset{A}{G}_g$. Then there is $\varphi \in \mathrm{X}(A)$ such that $\varphi|\overline{G}_g \equiv \chi$, $\varphi|D = \chi_f$. In fact, to see this it is enough to observe that χ_f is $\overline{G}_g$-invariant, $\overline{G}_g \cap D$ is connected and $\chi|(\overline{G}_g \cap D) \equiv \chi_f|(\overline{G}_g \cap D)$. Let B be equal to $\overline{G}_g \cdot L$. We define for $a \in A : \eta(a) \equiv \Delta_A(a)/\Delta_B(a)$ and consider, as in 4.1, $\mathcal{F} = \{f; f : B \to \mathbb{C}$ is measurable such that $|f|^2$ is locally integrable, $f(\delta x) = (\eta(\delta))^{1/2} \cdot \varphi(\delta) \cdot f(x)$ $(x \in B, \delta \in A)$ and $\mu_f(B/A)$ is finite$\}$.

Let H be the connected subgroup, corresponding to $\mathfrak{h}$, in $L_{\mathbb{C}}$. We note that there exists a continuous morphism $\chi_f : H \to \mathbb{C}^*$ such that $\chi_f(\exp(l)) \equiv \exp(-2\pi i(l,f))$ $(l \in \mathfrak{h})$. We put $e = (\mathfrak{h} \cap \bar{\mathfrak{h}}) \cap \mathcal{D}$ and write E for the corresponding subgroup of L. Let $a_0 \in B$ be fixed and F some element of $\mathcal{F}$. One verifies easily that if $h, h_1 \in H$ and $k, k_1 \in E$ are such that $hk = h_1 \cdot k_1$, then $\chi_f(h)F(k \cdot a_0) = \chi_f(h_1) \cdot F(k_1 \cdot a_0)$. Let us observe that since $\mathfrak{h} + e = e_{\mathbb{C}}$, HE is open in $E_{\mathbb{C}}$.

In this manner we can consider

$$\begin{aligned}\mathfrak{x} = \{F; F \in \mathcal{F}, &\text{such that for each } a_0 \in B, \text{ fixed,}\\ &\text{the map } hk \mapsto \chi_f(h)F(k \cdot a_0) \text{ is analytic on } HE\}.\end{aligned}$$

We put $U = \operatorname{ind}_{A\uparrow B}\varphi$. It is shown easily that the image T (say) of $\mathfrak{x}$ in $H(U)$ is closed and is invariant under the right-translations with elements of B. We define, for $p = (g, \chi)$, $\operatorname{ind}(\mathfrak{h}, p; B)$ as the restriction of U to T, to be quoted sometimes as the "holomorphic part" of U.

V. *Summary*

In a major part of the following, "holomorphic induction" will denote one of the following situations:

1) $\operatorname{ind}(\mathfrak{h}, p; B)$ as at the end of IV;
2) we write $\operatorname{ind}(\mathfrak{h}, p)$ for $\operatorname{ind}_{B\uparrow G}(\operatorname{ind}(\mathfrak{h}, p; B))$;
3) $\operatorname{Ind}(\mathfrak{h}, g)$ as to be explained next in VI.

Consider the disjoint union $\mathcal{R} = U_{g\in\mathfrak{g}^*}\overset{A}{G}_g$. Given $\chi \in \overset{A}{G}_g$, we define for a given $a \in G$, $a\chi \in \overset{A}{G}_{ag}$ as $(a\chi)(x) \equiv \chi(a^{-1}xa)$ $(x \in \overline{G}_{ag})$ and $a(g, \chi) = (ag, a\chi)$. In this fashion we have endowed $\mathcal{R}$ with the structure of a left G-space.

The main points of the ensuing discussion will be the following. For given $p \in \mathcal{R}$, $\operatorname{ind}(\mathfrak{h}, p)$ is a factor representation (4.5) independent, in the sense of unitary equivalence, of the particular choice of $\mathfrak{h} = \operatorname{pol}(p)$ (cf. 4.6). We denote by $F(p)$ the corresponding unitary equivalence class of unitary representations of G. We have $F(p) = F(p')$ if and only if the G-orbits of p and p', respectively, coincide (cf. 4.7).

Let us form $J(p) = \ker(F(p))$; these are members of $\operatorname{Prim}(G)$. The map $p \mapsto J(p)$ $(p \in \mathcal{R})$ is surjective (cf. 4.11). For given $J \in \operatorname{Prim}(G)$, its preimage in $\mathcal{R}$ will turn out to be a G-space, which is a foliation by coadjoint orbits of G ("generalized orbits"). The normal representation, corresponding to J, will be obtained by integration of members of $F(p)$, where p goes over the generalized orbit corresponding to J (cf. 4.12).

VI. *Review of the Kirillov theory*

For the following cf. [**AK**], Theorem I. Assume that $\mathfrak{g}$ is a nilpotent Lie algebra, and G the corresponding connected and simply connected Lie group. For $g \in \mathfrak{g}^*$, let $\mathfrak{h}$ be a polarization with respect to g. Let $\operatorname{Ind}(\mathfrak{h}, g)$ be the holomorphic part of $\operatorname{ind}_{D\uparrow G}\chi_f$. This being so we have the following results: $\operatorname{Ind}(\mathfrak{h}, g)$ is an irreducible unitary representation of g. It is independent of $\mathfrak{h}$ and also of g, as long as g moves along the same coadjoint orbit. Also, up to unitary equivalence, any irreducible unitary representation of G can be obtained in this form.

4.3. Extension of an irreducible representation

We recall that from now on $\mathfrak{g}$ is a finite-dimensional solvable Lie algebra, G the corresponding connected and simply connected Lie group. Also, we put $L = [G, G]$;

its Lie algebra, equal to $[\mathfrak{g}, \mathfrak{g}]$, will be denoted by $\mathcal{D}$. The objective of this section is the discussion of a result of crucial importance, due to Auslander and Kostant (cf. [**AK**], IV, 4.1), permitting the determination of the obstruction to extending a given irreducible representation of L to G.

LEMMA 1. *Let π be an irreducible unitary representation of L corresponding in the sense of Kirillov to the orbit Lf ($f \in \mathcal{D}^*$, fixed; cf. 4.2. VI). Then the stabilizer of π in G is equal to $G_f \cdot L$.*

In fact, let $\mathfrak{h}$ be a polarization with respect to f. With the notation of 4.2, $\operatorname{Ind}(\mathfrak{h}, f)$ is of the unitary equivalence class of π. Since, for any a in G, we have $a \operatorname{Ind}(\mathfrak{h}, f) = \operatorname{Ind}(a\mathfrak{h}, af)$, we have $a\pi = \pi$ if and only if $af \in Lf$ or $a \in G_f \cdot L$.

REMARKS. 1) Given $f \in \mathcal{D}^*$, let S be a closed submanifold of G_f such that $G_f = SL_f$ setwise. Given $a \in G_f$ we write $a = s(a)l(a)$ $(s(a) \in S, l(a) \in L_f)$. We can assume that the maps $a \mapsto s(a), a \mapsto l(a)$ are smooth. Note that we have also setwise $G_\pi = S \cdot L$.

2) Let f be a nonzero fixed element of $\mathcal{D}^*$ and $\mathfrak{h} = \operatorname{pol}(f)$ such that $G_f\mathfrak{h} = \mathfrak{h}$ (cf. II in 4.2). As is easily to be seen, there is a morphism ψ from G_π into the group of positive numbers, well determined by the conditions that if $a \in G_f$, then $\psi(a) = \det(\operatorname{Ad}(a^{-1})|\mathcal{D}/d)$ and $\psi|L \equiv 1$. Below dv will stand for an invariant measure on L/D. Let $\mathfrak{x}$ be the collection of all measurable complex-valued functions g on L such that $|g(x)|^2$ is locally integrable, $g(ax) \equiv \chi_f(a)g(x)$ $(a \in D, x \in L)$; we assume that for any $a_0 \in L$, fixed, $hk \mapsto \chi_f(h)g(ka_0)$ $(h \in H, k \in E)$ is holomorphic on HE and

$$\int_{L/D} |g(x)|^2 \cdot dv(p) < +\infty.$$

As explained earlier, this setup gives rise, through translation on the right, to an irreducible unitary representation of L. We continue to write $\mathfrak{x}$ for the representation space.

LEMMA 2. *Assume that $a \in G_\pi$ is fixed and $a = tl$ $(t \in S, l \in L)$. We claim that the map $g(x) \mapsto (\psi(a))^{1/2} \cdot g(t^{-1}xa)$ $(g \in \mathfrak{x})$ gives rise to a unitary operator on $\mathfrak{x}$.*

In fact, let us define $(\rho(a)g)(x) \equiv (\psi(a))^{1/2} \cdot g(t^{-1}xa)$.

(i) We note that if $\delta \in D, x \in L$, then

$$(\rho(a)g)(\delta x) = \chi_f(\delta)(\rho(a)g)(x).$$

In fact, we have

$$(\rho(a)g)(\delta x) = (\psi(a))^{1/2} \cdot g(t^{-1}\delta a) = (\psi(a))^{1/2} \cdot g(t^{-1}\delta \cdot t \cdot t^{-1}xa).$$

But, by $t \in G_f : \chi_f(t^{-1}\delta t) = \chi_f(\delta)$ and thus

$$(\rho(a)g)(\delta x) = \chi_f(\delta)(\rho(a)g)(x).$$

(ii) We note next that if $a_0 \in L$ is fixed, then the map $hk \mapsto \chi_f(h)(\rho(a)g)(ka_0)$ $(h \in H, k \in E)$ is holomorphic on HE.

In fact, we have by a previous reasoning,

$$\chi_f(h)g(t^{-1}ka_0a) = \chi_f(t^{-1}ht)g(t^{-1}kt \cdot t^{-1}a_0a),$$

whence it is enough to observe that the map $hk \mapsto t^{-1}hkt$ from HE into itself is holomorphic.

(iii) All this being so, it is enough to show that

$$\int_{L/D} |(\rho(a)g)(x)|^2 \cdot dv(p) = \int_{L/D} |g(x)|^2 \cdot dv(p).$$

To this end we can evidently assume that $a \in G_f$.

a) Since G_f normalizes D, G_f acts on the right on L/D by $Dx \mapsto D(a^{-1}xa)$.

b) We recall that for $a \in G_f$: $\psi(a) = \det(\mathrm{Ad}(a^{-1})|\mathcal{D}/d)$. We note that $dv(pa) = \psi(a) \cdot dv(p)$.

In fact, let h be continuous and of compact support on L. We have then

$$H(pa) = \int_D h(\delta a^{-1}xa) \cdot d\delta = \int_D h(a^{-1}(a\delta a^{-1})xa) \cdot d\delta.$$

This yields by the substitution $\delta \mapsto a^{-1}\delta a$

$$H(pa) = \det(\mathrm{Ad}(a^{-1})|d) \cdot \int_D h(a^{-1}\delta xa) \cdot d\delta$$

and thus, if dv is properly normalized

$$\begin{aligned}\int_{L/D} H(pa) \cdot dv(p) &= \det(\mathrm{Ad}(a^{-1})|d) \int_L h(a^{-1}xa)\, dx \\ &= \det(\mathrm{Ad}(a)) \det(\mathrm{Ad}(a^{-1})|d) \int_{L/D} H(p) \cdot dv(p)\end{aligned}$$

whence $dv(pa^{-1}) = \psi(a^{-1}) \cdot dv(p)$. Summing up, if $\psi(a) = \det(\mathrm{Ad}(a^{-1})|\mathcal{D}/d)$ $(a \in G_f)$, then we have $dv(pa) = \psi(a) \cdot dv(p)$.

c) From here we can conclude that

$$\int_{L/D} |(\rho(a)g)(x)|^2 \cdot dv(p) = \psi(a) \int_{L/D} |g(t^{-1}xt)|^2 \cdot dv(p) = \int_{L/D} |g(x)|^2 \cdot dv(p).$$

In this fashion, we have completed proving that if for $a = tl$ $(t \in S, l \in L)$ and for $g \in \mathfrak{x}$ we define

$$(\rho(a)g)(x) = (\psi(a))^{-1} g(t^{-1}xa)$$

where $\psi(a) = \det(\mathrm{Ad}(a^{-1})|\mathcal{D}/d)$, then $\rho(a)$ defines a unitary map of $\mathfrak{x}$ onto itself.

LEMMA 3. *With the above notation, there is a projective extension ρ of π to G_π such that if $a, b \in G_\pi$, $r = s(a), t = s(b)$, then*

$$\rho(a)\rho(b) = \omega(a,b)\rho(ab)$$

where $\omega(a,b) = \overline{\chi_f(l(rt))}$.

In fact, let g be some element of $\mathfrak{x}$; then, we have by definition, $(\rho(ab)g)(x) \equiv (\psi(ab))^{1/2} \cdot g(s(ab))^{-1} \cdot x \cdot ab)$. We have also if $a = s(a)m$, $b = s(b)n$ $(m, n \in L), r = s(a), t = s(b)$

$$ab = rm \cdot tn = rt(t^{-1}mt)n = s(rt)[l(rt) \cdot t^{-1}mt \cdot n]$$

whence $s(ab) = s(rt)$.

On the other hand, $(\rho(b)g)(x) \equiv (\psi(b))^{1/2} \cdot g(t^{-1}xb)$ and thus also

$$\rho(a)(\rho(b)g)(x) \equiv (\psi(ab))^{1/2} \cdot g(t^{-1}r^{-1} \cdot x \cdot ab).$$

In this manner it suffices to note that $rt = s(rt)l(rt)$ and thus $(rt)^{-1} = (l(rt))^{-1}(s(rt))^{-1}$. Summing up we have proved that if $a, b \in G_\pi$, then

$$\rho(a)\rho(b) = \omega(a,b)\rho(a)\rho(b)$$

where $\omega(a,b) = \overline{\chi_f(l(rt))}$.

4.4. Computation of U_π and K_π

The objective of this section is the computation for a given $\pi \in \widehat{L}$ of the groups $\mathcal{U}_\pi, K_\pi$ (cf. 3.1) in terms of the Kirillov orbit of π (we shall drop, in most cases, the π as an index). Let g be an element of $\mathfrak{g}^*$; we put $f = g|\mathcal{D}$, and assume $f \neq 0$. Let us write $B(x,y) \equiv ([x,y],f)$ $(x,y \in \mathfrak{g})$. If A is some subspace of $\mathfrak{g}$, we write $A_B^\perp$ for its orthogonal complement with respect to B. This being so we note that $\mathfrak{g}_f = \mathcal{D}_B^\perp$.

LEMMA 1. *With the above notation let us write* $k_\pi = \mathcal{D} + \mathfrak{g}_g$. *Then we have*

$$(G_f)_0 g = g + k_\pi^\perp.$$

In fact,

(i) We start with the observation that, evidently, if $n \geq 2$ and $\{l_1, \dots, l_n\}$ are elements in $\mathfrak{g}_f$, then $l_1 \cdots l_n g = 0$. This entails $(G_f)_0 g = g + \mathfrak{g}_f \cdot g$.

(ii) Since $\mathfrak{g}_f g$ is orthogonal to k_π, to attain our goal it is enough to verify that $\dim(\mathfrak{g}_f g) = \operatorname{codim}(k_\pi)$. In fact, we have evidently $\dim(\mathfrak{g}_f g) = \dim(\mathfrak{g}_f) - \dim(\mathfrak{g}_g)$. On the other hand,

$$\begin{aligned} \operatorname{codim}(k_\pi) &= \dim(\mathfrak{g}) - \dim(\mathcal{D}) - \dim(\mathfrak{g}_g) + \dim(\mathcal{D}_g) \\ &= \dim(\mathfrak{g}_f) - \dim(\mathfrak{g}_g). \end{aligned}$$

In this manner we have completed proving the following:

If g is some element of $\mathfrak{g}^*$, $f = g|\mathcal{D}$ and $k_\pi = \mathcal{D} + \mathfrak{g}_g$, then $(G_f)_0 \cdot g = g + k_\pi^\perp$.

LEMMA 2. *Suppose that* $g \in \mathfrak{g}^*$ *and* $u \in \mathcal{D}^\perp$. *We have then* $G_{g+u} = G_g$, $\overline{G}_{g+u} = \overline{G}_g$.

In fact,

(i) We start by observing that, evidently, if $a \in G, u \in \mathcal{D}^\perp$, then $au = u$. Hence, if $a \in G_g$, then $ag = g$ and $au = u$. In this manner we conclude that $a(g+u) = g+u$, whence $G_g \subseteqq G_{g+u}$. By symmetry, this implies that $G_{g+u} = G_g$.

(ii) We recall (cf. 4.2) that if $g \in \mathfrak{g}^*$, then

$$\overline{G}_g = \{a; a \in G_g \text{ such that } \chi_g([a,b]) = 1 \text{ for all } b \in G_g\}.$$

Since $u \in \mathcal{D}^\perp$ this implies that $\chi_g([a,b]) = \chi_{g+u}([a,b])$ and hence $\overline{G}_g = \overline{G}_{g+u}$, completing the proof of Lemma 2.

LEMMA 3. *Putting* $\mathfrak{g}_\pi = \mathcal{D} + \mathfrak{g}_f$, *we have* $L_f g = g + \mathfrak{g}_\pi^\perp$.

In fact,

(i) As in the proof of Lemma 1, we have if $\{l_1, \dots, l_n\} \subset \mathcal{D}_f$ and $n \geq 2$: $l_1 \cdots l_n g = 0$. This implies at once: $L_f g = g + \mathcal{D}_f g$.

(ii) We note next that $\mathfrak{g}_\pi$ is orthogonal to $\mathcal{D}_f g$.

To show this, it is enough to recall that $\mathfrak{g}_\pi = \mathcal{D} + \mathfrak{g}_f$, $\mathfrak{g}_f = \mathcal{D}_B^\perp$ and that, trivially, $\mathcal{D}_f g$ is orthogonal to $\mathcal{D}$ and to $\mathfrak{g}_f$.

(iii) To complete the proof of our lemma, it is enough to verify that $\dim(\mathcal{D}_f g) = \operatorname{codim}(\mathfrak{g}_\pi)$. In fact, we have $\dim(\mathcal{D}_f g) = \dim(\mathcal{D}_f) - \dim(\mathcal{D}_g)$. On the other hand,

$\operatorname{codim}(\mathfrak{g}_\pi) = \dim(\mathfrak{g}) - \dim(\mathfrak{g}_\pi)$ and thus

$$\begin{aligned}\operatorname{codim}(\mathfrak{g}_\pi) &= \dim(\mathfrak{g}) - \dim(\mathcal{D}) - \dim(\mathfrak{g}_f) + \dim(\mathcal{D}_f)\\ &= \dim(\mathfrak{g}) - \dim(\mathcal{D}) + \dim(\mathcal{D}_g) - \dim(\mathfrak{g}_f)\\ &\quad + \dim(\mathcal{D}_f) - \dim(\mathcal{D}_g) = \dim(\mathcal{D}_f) - \dim(\mathcal{D}_g)\end{aligned}$$

whence $\operatorname{codim}(\mathfrak{g}_\pi) = \dim(\mathcal{D}_f) - \dim(\mathcal{D}_g) = \dim(\mathcal{D}_f \cdot g)$.

Summing up, we have completed proving that if we define $\mathfrak{g}_\pi = \mathcal{D} + \mathfrak{g}_f$, then $L_f g = g + \mathfrak{g}_\pi^\perp$.

LEMMA 4. *Let us write $\mathcal{T} = \{a; a \in G_f$ such that $ag|\mathfrak{g}_f = g|\mathfrak{g}_f\}$. Then we have $\mathcal{T} = G_g \cdot L_f$.*

In fact, since the right-hand side is contained in G_f, it is enough to note that, by the proof of the previous lemma, if a is in L_f, then $ag = g +\ u$, where u is orthogonal to $\mathcal{D} + \mathfrak{g}_f$. In this manner $ag|\mathfrak{g}_f = g|\mathfrak{g}_f$. Let us assume now that $a \in G_f$ and $ag|\mathfrak{g}_f = g|\mathfrak{g}_f$. We have always $ag|\mathcal{D} = g|\mathcal{D}$. Hence, if $a \in G_f$, then $ag|(\mathcal{D} + \mathfrak{g}_f) = g(\mathcal{D} + \mathfrak{g}_f)$ and thus $ag = g + u$ where $u|(\mathcal{D} + \mathfrak{g}_f) = 0$. By Lemma 3, there is $l \in L_f$ such that $ag = lg$; therefore $a \in L_f G_g$, completing our proof of Lemma 4.

We recall that there is $\chi_f \in \mathrm{X}(L_f)$ such that $d\chi_f(l) = -2\pi i(l, f)$ $(l \in \mathcal{D}_f)$. We note that χ_f cannot be identically one. In fact, otherwise $f|\mathcal{D}_f = 0$. But, since $\mathcal{D}_f = (\mathcal{D}f)^\perp$, there is then an l in $\mathcal{D}$ such that $f = lf$. But this is impossible, since $\mathcal{D}$ is nilpotent and f is different from zero.

We note that $G_g \cap L_f = L_g$ and $\mathcal{D}_g \subseteqq \mathcal{D}_f$. We have also $\chi_g|L_g = \chi_f|L_g$. Let us put $\overset{\circ}{L}_f = \ker(\chi_f)$; clearly, it is invariant in G_f. In this manner we can form the central extension of G_f by the 1-torus $L_f/\overset{\circ}{L}_f$:

$$1 \longrightarrow L_f/\overset{\circ}{L}_f \xrightarrow{\text{central}} G_f/\overset{\circ}{L}_f \xrightarrow{\sigma} G_f/\overset{\circ}{L}_f \longrightarrow 1.$$

Below we shall write $\Lambda = G_f/\overset{\circ}{L}_f$.

LEMMA 5. *Let us write $\mathcal{T} = G_g \cdot L_f$.*
We have

$$\overline{G}_g L_f = \{a; a \in \mathcal{T} \text{ such that } [a,b] \in \ker(\chi_f) \text{ for all } b \text{ in } \mathcal{T}\}.$$

In fact, by what we saw above it is enough to observe that

$$\overline{G}_g = \{a; a \in G_f \text{ such that } \sigma([a,b]) = 1 \text{ for all } b \in G_f\}.$$

Below Γ will be as in Chapter II; for unexplained notation cf. loc. cit.

LEMMA 6. *We claim that $U^{\mathrm{c}} = \Gamma_1^{\mathrm{c}}$.*

In fact, we recall that $U = \Gamma_0 \cdot \Gamma_1$, and thus, if γ is in Γ_1^{c}, then $\gamma \in U^{\mathrm{c}}$. Conversely, if δ is in U^{c}, then δ commutes with Γ_0, whence δ is in Γ, and hence also in Γ_1^{c}.

LEMMA 7. *We claim that $U/\Gamma_0 \cdot U^{\mathrm{c}} = \Gamma_1/\Gamma_1^{\mathrm{c}}$.*

We have, in fact,

$$U/\Gamma_0 \cdot U^{\mathrm{c}} = \Gamma_0 \cdot \Gamma_1/\Gamma_0 \cdot U^{\mathrm{c}};$$

hence it is enough to show that $\Gamma_0 \cap \Gamma_1 = \Gamma_0 \cap \Gamma_1^{\mathrm{c}}$, which is true by $\Gamma_1 \cap \Gamma_0 \subset \Gamma_1^{\mathrm{c}}$.

Summing up, we have shown that $U^c = \Gamma_1^c$.

LEMMA 8. *Assume that $Z = \mathbb{R}^a \times \mathbb{Z}^b$ and that Γ is a central extension, by a 1-torus, of Z. Let us write $\gamma = \mathrm{Lie}(\Gamma)$ and e for an element, generating the 1-torus in Γ. If $d \in \gamma^*$ is such that $(e, d) \neq 0$, then $\Gamma_d = \Gamma_1$.*

In fact, we recall that there is a skew-symmetric real, bilinear form B on $Z \times Z$, such that if we put $\alpha(a, b) \equiv \exp((i/2)(B(a, b))$, then

$$\Gamma = \{(a, u); a \in Z, u \in T \text{ such that } \\ (a, u)(b, v) = (a + b, \alpha(a, b)uv)\}.$$

An easy computation shows that if c is a fixed real number, then

$$(a, u)(t \cdot z, \exp(ict))(a, u)^{-1} = (tz, \exp(i(B(a, z) + c)t)).$$

We have thus obtained $\mathrm{Ad}((a, u))(z, c) = (z, B(a, z) + c)$. Let us put $\gamma = Z_0 + \mathbb{R}e$ and correspondingly $\gamma^* = Z_0^* + \mathbb{R}e^*$. We define for $a \in Z$ fixed: $F(a) \in Z_0^*$ by $F(a)(z) \equiv B(a, z)$. Then we have for any $z \in Z_0$, $k, k' \in \mathbb{R}$,

$$\mathrm{Ad}((a, u))(z + ke) = z + (kt + B(a, z))e$$

and

$$(\mathrm{Ad}(a, u))'(z' + k'e^*) = (z' + k'F(a)) + k'e^*.$$

We conclude from here that if $k' \neq 0$ is fixed, then we have

$$(\mathrm{Ad}(a, u))'(z' + k'e^*) = z' + k'e^*$$

if and only if $F(a) = 0$.

Let π be an element of $\widehat{L}$, and $Lf \subset \mathcal{D}^*$ the corresponding coadjoint orbit. As we saw in 4.3, there is a projective extension ρ of π to G_π such that

$$\rho(a)\rho(b) = \omega(a, b)\rho(ab)$$

where, if $a, b \in G_\pi$ and $r = s(a)$, $t = s(b)$, then

$$\omega(a, b) = \overline{\chi_f(l(rt))}.$$

We have evidently $\omega(am, bn) = \omega(a, b)$ $(m, n \in L)$. We form the Mackey group M_π by

$$\phi : G_\pi^e \longrightarrow M_\pi = G_\pi^e / L.$$

We shall write Γ for M_π. We conclude from all this that Γ can be realized as

$$\{(r, u) : r \in S, u \in T, \text{ s.t. } (r, u)(t, v) = (s(rt), \overline{\chi_f(l(rt))}uv).$$

We define the map $\psi : G_f \to \Gamma$ by $\psi(a) = (s(a), \overline{\chi_f(l(a))})$.

LEMMA 9. *With the above notation, ψ is a surjective morphism of groups such that $\ker(\psi) = \overset{\circ}{L}_f$ $(= \ker(\chi_f))$.*

In fact,

(i) To show that ψ is a morphism of groups $G_f \to \Gamma$, assume that $a = rm, b = tn$ $(r, t \in S; m, n \in L_f)$. We have

$$\begin{aligned}\psi(a)\psi(b) &= (r, \overline{\chi_f(m)})(t, \overline{\chi_f(n)}) \\ &= (s(rt), \overline{\chi_f(l(rt))}\,\overline{\chi_f(m)}\,\overline{\chi_f(n)}).\end{aligned}$$

We have also

$$\begin{aligned}\psi(ab) &= \psi(rm \cdot tn) = \psi(s(rt)[l(rt) \cdot t^{-1}mt \cdot n]) \\ &= (s(rt), \overline{\chi_f(l(rt) \cdot t^{-1}mt \cdot n)}) \\ &= (s(rt), \overline{\chi_f(l(rt))} \cdot \overline{\chi_f(t^{-1}mt)} \cdot \overline{\chi_f(n)}).\end{aligned}$$

But since $t \in S$, we have $\chi_f(t^{-1}mt) = \chi_f(m)$ and thus

$$\psi(ab) = (s(rt), \overline{\chi_f(l(rt))} \cdot \overline{\chi_f(m)} \cdot \overline{\chi_f(n)})$$

whence we can conclude that

$$\psi(ab) = \psi(a)\psi(b) \qquad (a, b \in G_f).$$

(ii) We observe that, since $\psi(rm) = (r, \overline{\chi_f(m)})$, we have $\ker(\psi) = \overset{\circ}{L}_f$.

(iii) Finally we note that ψ is surjective. In fact, as we saw above, by $f \neq 0$ we have $f|\mathcal{D}_f \neq 0$ and thus $\chi_f|L_f \neq 1$.

Summing up, we have proved that if we define for $a \in G_f$, $\psi(a) = (s(a), \overline{\chi_f(l(a))})$, then ψ is a surjective morphism $G_f \to \Gamma$ such that $\ker(\psi) = \overset{\circ}{L}_f$.

We recall that $\sigma : G_f \to G_f/\overset{\circ}{L}_f = \Lambda$.

We write $j : \Gamma \to \Lambda$ such that

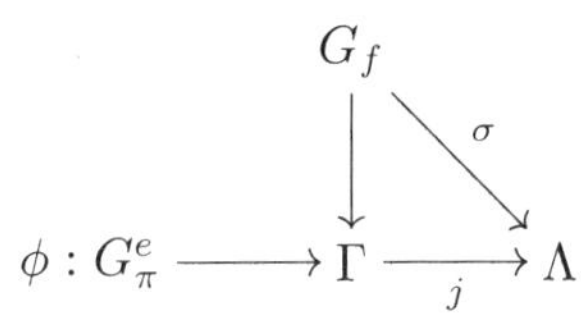

We write $\lambda = \mathrm{Lie}(\Lambda)$, and $\overset{\circ}{\mathcal{D}}_f = \mathrm{Lie}(\overset{\circ}{L}_f)$. There is evidently an element α in λ^* such that $(\delta\sigma)(\alpha) = g|\mathfrak{g}_f$. We write w for an element in λ such that $\exp(\mathbb{R} \cdot w) = L_f/\overset{\circ}{L}_f$. We claim that $(w, \alpha) \neq 0$. In fact, by $\chi_f|L_f \not\equiv 1$, there is $l \in \mathcal{D}_f - \overset{\circ}{\mathcal{D}}_f$ such that $\alpha\sigma(l) = w$, and we have

$$(w, \alpha) = (\alpha\sigma(l), \alpha) = (l, (\delta\sigma)(\alpha)) = (l, g) \neq 0.$$

Notation. Below we use notation, analogous to that employed in connection with Γ, with respect to Λ. In particular, Lemma 8 implies $\Lambda_d = \Lambda_1$.

We recall that by Lemma 4 we have

$$G_g \cdot L_f = \{a; a \in G_f \text{ such that } ag|\mathfrak{g}_f = g|\mathfrak{g}_f\}.$$

This implies

LEMMA 10. *We have* $G_g L_f = \sigma^{-1}(\Lambda_d)$.

Let us recall that by Lemma 5

$$\begin{aligned}\overline{G}_g L_f = \{a; a \in G_f L_f, \chi_f([a, b]) \equiv 1 \\ \text{for any } b \in G_g L_f\}.\end{aligned}$$

From this we infer

LEMMA 11. *We have* $\overline{G}_g L_f = \sigma^{-1}(\Lambda_1^e)$.

We recall that (cf. Lemma 6 in 3.1) $K_\pi^e = \phi^{-1}(U^c)$. From this we get

LEMMA 12. *We claim that* $K_\pi = \overline{G}_g L$.

In fact, we recall that by Lemma 6 : $U^c = \Gamma_1^c$. Lemma 8 implies that $\Lambda_1 = \Lambda_d$ and Lemma 11 that $\overline{G}_g \cdot L_f = \sigma^{-1}(\Lambda_d^c)$. Hence it suffices to recall that

$$\begin{array}{ccccc} & & G_f & & \\ & & \downarrow \psi & \searrow \sigma & \\ \phi : G_\pi^e & \longrightarrow & \Gamma & \xrightarrow{j} & \Lambda \end{array}$$

We continue with the notation of Lemma 6 in 3.1. Hence $\mathcal{U}^0 = \phi^{-1}(U)$.

LEMMA 13. *We claim that*

$$\mathcal{U}/(G_\pi)_0 \cdot K = G_g/\overline{G}_g.$$

In fact, we recall that

$$\mathcal{U}^e = \phi^{-1}(U), \qquad (G_\pi)_0^e = \phi^{-1}(\Gamma_0), \qquad K_\pi^e = \phi^{-1}(U^c).$$

We have by Lemma 7 and by what we saw above

$$U/\Gamma_0 U^c = \Gamma_1/\Gamma_1^c = G_g/\overline{G}_g$$

and hence

$$\mathcal{U}/(G_\pi)_0 \cdot K = G_g/\overline{G}_g.$$

We note finally

LEMMA 14. *We claim that*

$$\mathcal{U}_\pi = (G_f)_0 \cdot G_g \cdot L.$$

In fact, to see this, it suffices to recall $U = \Gamma_1 \cdot \Gamma_0$ and that

$$\Gamma_1 = \psi(\sigma^{-1}(\Lambda_d)) = \psi(G_g L_f)$$

and

$$\Gamma_0 = \psi(\sigma^{-1}(\Lambda_0)) = \psi((G_f)_0).$$

4.5. Holomorphically induced representations

Let π in $\widehat{L}$ be given. We write $\mathfrak{G}(\pi) = \{\lambda; \lambda \in \widehat{K}_\pi \text{ such that } \lambda|L = \pi\}$. Our objective in this section will be to describe elements of $\mathfrak{G}(\pi)$ as holomorphically induced representations. Let g be some element of $\mathfrak{g}^*$ such that, if $f = g|\mathcal{D}$, then π belongs to the orbit Lf. We shall write B for K_π; recall that $B = \overline{G}_g L$ (cf. Lemma 12). Let $\mathfrak{h}$ be a polarization with respect to f such that $\overline{G}_g \mathfrak{h} = \mathfrak{h}$. Writing $d = \mathfrak{h} \cap \mathcal{D}$, d is invariant with respect to $\overline{G}_g$. The morphism $a \mapsto \det(\mathrm{Ad}(a^{-1})|\mathcal{D}/d)$ $(a \in \overline{G}_g)$ admits an extension ψ, which is identically one on L, to B. Let $\mathfrak{x}$ have a meaning analogous to the one used in 4.3. Let $\overline{G}_g = SL_g$ similarly as loc. cit. We recall that if $a \in B$ such that $a = tm$ $(t \in S, m \in L)$ and $u \in \mathfrak{x}$, and if $(\rho(a)u)(x) \equiv (\psi(a))^{1/2} \cdot u(t^{-1}xa)$, then the map $a \mapsto \rho(a)$ defines a continuous projective extension of π to B. If $a, b \in B$ are such that $r = s(a)$, $t = s(b)$, then

$$\rho(a)\rho(b) = \overline{\chi_f(l(rt))} \cdot \rho(ab).$$

LEMMA 1. *With unexplained notation as above, let* $\chi \in \overset{A}{G_g}$ *be fixed and set* $\rho_\chi(a) \equiv \chi(s(a))\rho(a)$ $(a \in B)$. *We claim that* ρ_χ *is a unitary representation of* B *on* $\mathfrak{x}$.

In fact, if $a, b \in B$, let us put $r = s(a), t = s(b)$. Then $\rho_\chi(a)\rho_\chi(b) = \chi(rt)\rho(a)\rho(b)$. Since $rt = s(rt)l(rt)$ it is enough to observe that if $a = rm, b = tn$ $(r, t \in S; m, n \in L)$ then $s(ab) = s(rt)$. In fact, we have

$$ab = rm \cdot tn = rt(t^{-1}mt)n = s(rt)[l(rt) \cdot t^{-1}mt \cdot n]$$

and thus the desired conclusion follows by noting that the first factor in the right-hand side lies in S and the second in L.

Let $p = (g, \chi)$ be some element of $\mathcal{R}$. Below $\mathrm{ind}(\rho, p; B)$ will have the same meaning as in 4.2, IV.

LEMMA 2. *We claim that* ρ_χ *and* $\mathrm{ind}(\rho, p; B)$ *are unitarily equivalent.*

In fact,

(i) We set $A = \overline{G}_g \cdot D$ and $B = \overline{G}_g \cdot L$; then $\mathrm{Lie}(A) = d + \mathfrak{g}_g$ and $\mathrm{Lie}(B) = \mathcal{D} + \mathfrak{g}_g$. We have, if $a \in \overline{G}_g$:

$$\eta(a) \equiv \Delta_A(a)/\Delta_B(a) \equiv \det(\mathrm{Ad}(a^{-1})|\mathcal{D}/d) = \psi(a).$$

From here we conclude (cf. 4.1) that there is a Borel measure dw on B/A such that $dw(gk) = \psi(k) \cdot dw(g)$ $(k \in B)$. If $\mathfrak{x}'$ is the space of $\mathrm{ind}(\rho, p; B)$ and $g \in \mathfrak{x}'$, then

$$\mu_g(B/A) = \int_{B/A} (|g(x)|^2/\psi(x)) \cdot dw(g).$$

(ii) Let again $\overline{G}_g = SL_g$ setwise; then we have also $B = S \cdot L$ setwise. If $v \in \mathfrak{x}$, we define $V(x)$ on B by

$$V(x) \equiv (\psi(t))^{1/2} \cdot \chi(t) \cdot v(y) \quad (x = ty; t \in S, y \in L).$$

To attain our goal, it is enough to establish the following points.

I. We have for any $v \in \mathfrak{x} : V \in \mathfrak{x}'$ and, if $dw(g)$ is properly normalized, then

$$\mu_V(B/A) = \mu_v(L/D).$$

II. Let u be in $\mathfrak{x}$, a in B and let us write

$$v(y) \equiv (\rho_\chi(a)u)(y) \quad (y \in L).$$

Then we have $V(x) \equiv v(xa)$ on B.

In fact,

Proof of I.

1) Recalling that $A = \overline{G}_g \cdot D$, let us define $\varphi \in \mathrm{X}(A)$ by

$$\varphi|\overline{G}_g \equiv \chi|\overline{G}_g \text{ and } \varphi|D \equiv \chi_f.$$

As above, we set

$$U(ty) \equiv (\psi(t))^{1/2} \cdot \chi(t) \cdot u(y) \quad (t \in S, y \in L).$$

We claim that if $c \in A$ and $x \in B$, then

$$U(cx) \equiv (\psi(c))^{1/2} \cdot \varphi(c) \cdot U(x).$$

In fact, let us write $c = rm, x = ty$ $(r, t \in S, m \in D, y \in L)$. We have then:

$$cx = rm \cdot ty = s(rt)[l(rt) \cdot t^{-1}mt \cdot y]$$

and thus

$$U(cx) = (\psi(s(rt)))^{1/2} \cdot \chi(s(rt)) \cdot u(l(rt) \cdot t^{-1}mt \cdot y).$$

We have by $u \in \mathfrak{x}$

$$u(l(rt) \cdot t^{-1}mt \cdot y) = \chi_f(l(rt)) \cdot \chi_f(m) \cdot u(y)$$

whence

$$U(x) = (\psi(r))^{1/2} \cdot (\psi(t))^{1/2} \chi(s(rt)) \cdot \chi_f(l(rt)) \cdot \chi_f(m) \cdot u(y).$$

But $\varphi(c) = \chi(r) \cdot \chi_f(m), \psi(r) = \psi(c)$ and thus

$$U(cx) \equiv (\psi(c))^{1/2} \cdot \varphi(c) \cdot (\psi(t))^{1/2} \cdot \chi(t) \cdot u(y)$$

which is the same as

$$U(cx) \equiv (\psi(c))^{1/2} \cdot \varphi(c) \cdot U(x).$$

2) We observe next that if $a_0 \in B$ is fixed, then the map $hk \mapsto \chi_f(h)U(ka_0)$ $(h \in H, k \in E)$ is holomorphic on HE (open in $E_{\mathbb{C}}$). In fact, let us assume that $a_0 = r \cdot m$ $(r \in S, m \in L)$. We have then $ka_0 = r(r^{-1}kr) \cdot m$ and $U(ka_0) = \psi(r)^{1/2} \cdot \chi(r) \cdot u(r^{-1}kr \cdot m)$; hence

$$\chi_f(h)U(ka_0) = (\psi(r))^{1/2} \cdot \chi(r) \cdot \chi_f(r^{-1}kr) \cdot u(r^{-1}kr \cdot m)$$

proving that $hk \mapsto \chi_f(h)U(ka_0)$ is holomorphic on HE.

To complete the verification of I, it is enough to show that

3) If $dv(p)$ is invariant on L/D, and if $dw(g)$ is a Borel measure on B/A such that $dw(gk) = \psi(k)\, dw(g)$ then in view of 4.1

$$\int_{L/D} |g(y)|^2 \cdot dv(p) = \int_{B/A} (|g(x)|^2/\psi(x)) \cdot dw(g).$$

In fact, we note first that $\overline{G}_g$ operates on the right on L/D by

$$b(Dx)b^{-1} = D(b^{-1}xb).$$

We have by 4.3: $dv(pb) = \psi(b) \cdot dv(p)$. If we define $\gamma : L/D \to B/A$ by $\gamma(Dx) = Ax$, then we obtain $\gamma(pb) = \gamma(p)b$ $(b \in \overline{G}_g)$ and $\gamma(pl) = \gamma(p)l$ $(l \in L)$. From here we conclude that if $v' = \gamma_*(v)$, then $dv'(gk) = \psi(k) \cdot dv'(p)$ $(k \in B)$ implying that $v' = \text{const} \cdot w$ and

$$\int_{L/D} |g(y)|^2 \cdot dv(p) = \int_{B/A} (|g(x)|^2/\psi(x)) \cdot dw(g)$$

as claimed above. In this manner, to complete our proof of Lemma 2, it is enough to establish that

Proof of II. If $a = rm$ $(r \in S, m \in L)$ is fixed in B and $v = \rho_\chi(a)u$ (which says that $v(y) \equiv \chi(r)(\psi(a))^{1/2} \cdot u(r^{-1}ya), y \in L)$ then $V(x) \equiv U(xa)$ on B.

In fact, if $a = rm$, $x = ty$ $(r, t \in S, m, y \in L)$ then $xa = s(tr)(l(tr) \cdot r^{-1}ya)$, implying that

$$\begin{aligned} U(xa) &= (\psi(s(tr))^{1/2} \cdot \chi(s(tr)) \cdot u(l(tr) \cdot r^{-1}ya) \\ &= (\psi(t))^{1/2}\chi(t)\chi(r)(\psi(a))^{1/2} \cdot u(r^{-1}yu) \\ &= (\psi(t))^{1/2} \cdot \chi(t) \cdot v(x) = V(x) \end{aligned}$$

which completes the proof of our lemma.

4.6. Proof that ind is independent of the polarization

The basic reference for what follows is [**AK**], Theorem III.3.1.

Our objective in this section will be to show that $\operatorname{ind}(\mathfrak{h}, p; k)$, up to unitary equivalence, is independent of the choice of $\mathfrak{h} = \operatorname{pol}(p), \overline{G}_g\mathfrak{h} = \mathfrak{h}$.

To achieve this, we shall be repeating some previous computations to prepare the groundwork to invoke the relevant results of Auslander and Kostant.

We make a choice of (g, χ) in $\mathcal{R}$ and recall that if π corresponds to Lf, then $K_\pi = \overline{G}_g \cdot L$ (cf. 4.4, Lemma 12).

Below $\mathfrak{x}$ and ψ will have the meaning as in 4.3. We define

$$M = \{(b, l); b \in \overline{G}_g, l \in L \text{ such that } (b, l)(b_1, l_1) = (b \cdot b_1, b_1^{-1} l b_1 \cdot l_1)\}.$$

Also if $m = (b, l) \in M$ and $f \in \mathfrak{x}$, then

$$(V(m)g)(x) \equiv (\psi(b))^{1/2} \cdot g(b^{-1} x b \cdot l).$$

LEMMA 1. *With the above notation, V gives rise to a unitary representation of M on $\mathfrak{x}$.*

In fact,

(i) Assume that $m_1, m_2 \in M$. Then $V(m_1)V(m_2) = V(m_1 \cdot m_2)$. To show this, let us assume that $m_1 = (b, l), m_2 = (b', l')$ and that g is in $\mathfrak{x}$. We have then

$$(V(m_2)g)(x) = (\psi(b'))^{1/2} \cdot g(b'^{-1} x b' \cdot l')$$

implying that

$$\begin{aligned} V(m_1)V(m_2)g(x) &= (\psi(b))^{1/2}(\psi(b'))^{1/2} \cdot g(b'^{-1}b^{-1} \cdot x \cdot bl \cdot b'l') \\ &= (\psi(bb'))^{1/2} \cdot g((b \cdot b'^{-1})^{-1}(b \cdot b')x \cdot bb' \cdot b'^{-1} \cdot l \cdot b'l') \\ &= (V(m_1 m_2)g)(x). \end{aligned}$$

(ii) We claim that if $m \in M$ ($m = (b, l)$, say) and $g \in \mathfrak{x}$, then $V(m)g$ belongs to $\mathfrak{x}$. In fact, let δ be some element of D. Then $(V(m)g)(\delta x) = \chi_f(\delta)(V(m)g)(x)$. To show this we remark that $(V(m)g)(\delta x) = (\psi(b))^{1/2} \cdot g(b^{-1} \cdot \delta x \cdot bl)$. But

$$b^{-1} \cdot \delta x \cdot bl = b^{-1} \cdot \delta b \cdot b^{-1} x bl.$$

Writing $\delta' = b^{-1} \cdot \delta b$, we obtain

$$\begin{aligned} (V(m)g)(\delta x) &= (\psi(b))^{1/2} \cdot g(\delta' \cdot b^{-1} x bl) \\ &= \chi_f(\delta')(V(m)g)(x). \end{aligned}$$

Hence it is enough to note that if $\delta' = b^{-1}\delta b$ and $b \in \overline{G}_g$, then $\chi_f(\delta') = \chi_f(\delta)$. In this manner we can conclude that $(V(m)g)(\delta x) = \chi_f(\delta)(V(m)g)(x)$. We note next that if $h \in H, k \in E$ and a_0 lie in L then $\chi_f(h)\ (V(m)g)(ka_0)$ depends on hk only. To see this, assume that $hk = h'k'$. Then there is a δ in D such that $h' = h \cdot \delta^{-1}, k' = \delta k$. Hence by what we saw above,

$$\chi_f(h')(V(m)g)(k' \cdot a_0) = \chi_f(h)(V(m)g)(ka_0)$$

proving our claim.

Let us prove next that if $a_0 \in L$, $h \in H, k \in E$, then the map

$$hk \mapsto \chi_f(h)(V(m)g)(ka_0)$$

is holomorphic on HE. In fact, we have

$$\chi_f(h)(V(m)g)(ka_0) \equiv (\psi(b))^{1/2} \cdot \chi_f(h) \cdot g(b^{-1} k a_0 bl)$$

which is equal, by $b \in \overline{G}_g$, if we put $b^{-1}a_0bl = c$, to

$$(\psi(b))^{1/2}\chi_f(b^{-1}hb)g(b^{-1}kb \cdot c).$$

Hence it suffices to recall that, by $g \in \mathfrak{x}$, if c is in L, then $hk \mapsto \chi_f(h)g(kc)$ is holomorphic on HE. In this manner we have completed proving that if $g \in \mathfrak{x}$, $a_0 \in L, m \in M$, fixed, then the map

$$hk \mapsto \chi_f(h)(V(m)g)(ka_0)$$

is holomorphic on HE.

Let us prove finally that if $g \in \mathfrak{x}$ and $m \in M$ are fixed, then $\mu_{V(m)g}(L/D) = \mu_g(L/D)$. In fact, we recall that if $b \in \overline{G}_g$, then $dv(pb) = \psi(b) \cdot dv(p)$. We obtain in this fashion

$$\begin{aligned}(\mu_{V(m)g})(L/D) &= \int_{L/D} |V(m)g(x)|^2 \cdot dv(p)\\ &= \psi(b)\int_{L/D} |g(b^{-1}xbl)|^2 \cdot dv(p) = \psi(b)\int_{L/D} |g(xl)|^2 \cdot dv(p \cdot b^{-1})\\ &= \int_{L/D} |g(xl)|^2 \cdot dv(p) = \int_{L/D} |g(x)|^2 \cdot dv(p) = \mu_g(L/D)\end{aligned}$$

and hence $\mu_{V(m)g}(L/D) = \mu_g(L/D)$.

Summing up, we have completed proving Lemma 1 saying that V gives rise to a unitary representation of M on $\mathfrak{x}$.

LEMMA 2. *With the above notation, up to unitary equivalence, V is independent of the polarization employed in its construction.*

For this see [**AK**], Theorem III.3.1.

LEMMA 3. *The projective representation in Lemma 3 (4.3) and the unitary representations in Lemmas 1–2 (4.5), up to unitary equivalence, are independent of the polarization $\mathfrak{h}$ (subject to the condition of $\overline{G}_g$-invariance).*

In fact, for this it is enough to observe that, with the previous notation, if $a \in K$ is such that $a = rm$ ($r \in S, m \in L$), then $\rho(a) = V(r, m)$.

4.7. Condition for unitary equivalence

We recall that $\lambda \in \overset{A}{G}_g$ may stand for the pair (g, χ). We write

$$\mathcal{R} = \bigcup_{g \in \mathfrak{g}^*} \overset{A}{G}_g, \text{ disjoint union.}$$

We recall also that if χ is some element of $\overset{A}{G}_g$ and a is fixed in G, then we define $ax \in \overset{A}{G}_{ag}$ such that $(a\chi)(b) = \chi(a^{-1}ba)$ $(b \in \overline{G}_{ag})$.

Let us recall that if $p \in \mathcal{R}, \mathfrak{h} = \mathrm{pol}(p)$, then $\mathrm{ind}(\mathfrak{h}, p)$ is defined by $\mathrm{ind}_{K\uparrow G}(\mathrm{ind}(\mathfrak{h}, p; K))$. By 4.6 it is independent of $\mathfrak{h} = \mathrm{pol}(p)$.

We write $F(p)$ for its unitary equivalence class: By Lemma 9 in 3.1, $F(p)$ is composed of semi-finite factor representations. By the proof of Lemma 10 in 3.1, if members of $F(p_1)$ and $F(p_2)$ resp. are quasi-equivalent, then $F(p_1) = F(p_2)$. By definition, evidently, if $p_2 = ap_1$, then $F(p_1) = F(p_2)$.

Our objective below will be to establish the opposite implication: If $F(p_1) = F(p_2)$, then $Gp_1 = Gp_2$. In fact,

(i) Lemma 10 in 3.1 implies that there is $a \in G$ such that $a\,\mathrm{ind}(\mathfrak{h}_1, p_1; K) = \mathrm{ind}(\mathfrak{h}_2, p_2; K)$. Since for any $a \in G, p \in \mathcal{R}, \mathfrak{h} = \mathrm{pol}(p)$

$$a\,\mathrm{ind}(\mathfrak{h}, p; K) = \mathrm{ind}(a\mathfrak{h}, ap; K)$$

we can assume that $\mathrm{ind}(\mathfrak{h}_1, p_1; K) = \mathrm{ind}(\mathfrak{h}_2, p_2; K)$ and it suffices to derive from this that there is $a \in G$ such that $p_2 = ap_1$.

(ii) By 4.6, the unitary equivalence class of $\mathrm{ind}(\rho, p; K)$ is independent of $\mathfrak{h}$; we denote it by $\lambda(p)$. Thus it suffices to show that if $\lambda(p_1) = \lambda(p_2)$, then the G-orbits of p_1 and p_2 coincide.

(iii) Evidently, given $p = (g, \chi)$, the restriction of $\mathrm{ind}(\mathfrak{h}, p; K)$ to L is an irreducible representation. If $f = g|\mathcal{D}$, it corresponds, in the sense of Kirillov, to the orbit $Lf \subset \mathcal{D}^*$. Thus we can assume from the start that $g_1|\mathcal{D} = f = g_2|\mathcal{D}$ and hence, by Lemma 1 in 4.4, $\overline{G}_{g_1} = \overline{G}_{g_2}, = C$ (say).

(iv) We assume $\mathrm{ind}(\mathfrak{h}, p_1; K) = \mathrm{ind}(\mathfrak{h}, p_2, K)$. By Lemma 2 in 4.5, with notation as loc. cit., $\mathrm{ind}(\mathfrak{h}, g, \chi; K)$ is equivalent to ρ_2, and thus ρ_{χ_1} is equivalent to ρ_{χ_2}, implying $\chi_1 = \chi_2, = \chi$, say. We put $\mathfrak{a} = \mathrm{Lie}(C)$ and $k_\pi = \mathcal{D} + \mathfrak{a}$. Hence $g_1|k_\pi = g_2|k_\pi$, and it is enough to verify that if $(g_1, \chi), (g_2, \chi)$ are given such that $g_1|k_\pi = g_2|k_\pi$, then there is $a \in G_f$ such that $ag_1 = g_2$ (cf. Lemma 1 in 4.4) and $a\chi = \chi$.

(v) By the end of 4.4, we have $\sigma((G_f)_0) = \Lambda_0$ and $\sigma(\overline{G}_g L_f) = \Lambda_1^c$, in which fashion $\chi_g(aba^{-1}b^{-1}) \equiv 1$ ($a \in (G_f)_0$, fixed; $b \in \overline{G}_g$) implying $a\chi = \chi$.

As a corollary of the above proof, we obtain that if we put $M = G_g \cdot (G_f)_0 \cdot L$, and ρ is fixed in F, then $\lambda^{-1}(\rho)$ is an M-orbit.

4.8. Regularized orbits

The objective of the next two sections is the introduction for an arbitrary connected and simply connected solvable Lie group G with the Lie algebra $\mathfrak{g}$, of a series of geometric objects, which we call generalized orbits. These will turn out to be in a natural bijection with the underlying set of $\mathrm{Prim}(G)$. For a nilpotent G, generalized orbits coincide with coadjoint orbits. In the general case, a generalized orbit is a torus bundle over a subset of $\mathfrak{g}^*$; the latter, to be introduced next, is what we call a regularized orbit or R-orbit for short. Their definition takes place due to the following.

PROPOSITION. *With the previous notation there is an equivalence relation, R, say, on $\mathfrak{g}^*$, well-determined by the following properties.*

1) *An R-orbit, O, say, is invariant with respect to the coadjoint action of G on $\mathfrak{g}^*$, containing any of its G-orbits as dense subsets.*

2) *O is locally closed in $\mathfrak{g}^*$.*

Let us observe that the above proposition works out in much more general situations. In particular, it is valid for any connected Lie group (solvability dropped). As pointed out by J. M. Rosenberg, the R-relation can also be defined as the finest G-invariant equivalence relation on $\mathfrak{g}^*$, for which the corresponding orbit space is T_0.

UNIQUENESS. To this end it is enough to show that if O_1 and O_2 are as O above, such that their intersection is nonempty, then $O_1 = O_2$. In fact, we start by noting that if $E \subset \mathfrak{g}^*$ is G-invariant, such that $E \subset \overline{Gx}$ for some x in E, then

$\overline{E} = \overline{Gx}$. This implies at once that the closures of O_1 and O_2 coincide; we shall denote it by F. Let us assume, say, that x is contained in $O_1 - O_2$. The latter is a subset of $F - O_2$. Since by assumption, O_2 is open in its closure, it is closed; thus it contains $\overline{Gx}$. But $O_2 \subseteqq \overline{Gx}$, implying a contradiction.

EXISTENCE. We observe that there is an equivalence relation, $\sim$ say, on $\mathfrak{g}^*$ such that $x \sim y$ if and only if $\overline{Gx} = \overline{Gy}$. Let H be an equivalence class with respect to $\sim$. We claim that H is locally closed in $\mathfrak{g}^*$. In fact

a) We can assume to have the following situation: $\tilde{\mathfrak{g}}$ is an algebraic Lie algebra, such that $\tilde{\mathfrak{g}} \supseteqq \mathfrak{g}$ and $[\tilde{\mathfrak{g}}, \tilde{\mathfrak{g}}] = [\mathfrak{g}, \mathfrak{g}]$. We denote by $\widetilde{G}$ the connected and simply connected Lie group, such that $\mathrm{Lie}(\widetilde{G}) = \tilde{\mathfrak{g}}$. Since $\mathfrak{g}$ is an ideal in $\tilde{\mathfrak{g}}$, $\widetilde{G}$ acts on $\mathfrak{g}^*$, and the $\widetilde{G}$-orbits are locally closed.

b) We claim that H is contained in a $\widetilde{G}$-orbit. In fact, if not, then there are $\widetilde{G}$-orbits $O_1, O_2 \subset \mathfrak{g}^*$ such that $O_1 \neq O_2$ and $H \cap O_j \neq 0$ $(j = 1, 2)$. If O is any $\widetilde{G}$-orbit such that $O \cap H \neq 0$, we can write $O = \mathscr{O} \cap \mathcal{F}$, where $\mathscr{O}$ is open, $\mathcal{F}$ closed; both can be assumed to be $\widetilde{G}$-invariant. If x is in $\mathcal{F} \cap H$, then $\overline{H} \subseteqq \overline{Gx} \subseteqq \mathcal{F}$; hence $O \cap \overline{H} = \mathscr{O} \cap \overline{H}$ is open and, evidently, dense in $\overline{H}$. Therefore $O_j \cap \overline{H}$ both are open and dense in $\overline{H}$ and they are disjoint. The contradiction so arising proves that H is contained in a $\widetilde{G}$-orbit.

c) 1) Assume that O is a $\widetilde{G}$-orbit. We claim that if x is in O, then $G \cdot \widetilde{G}_x$ is invariant in $\widetilde{G}$. In fact, to this end it is enough to note that $\widetilde{G}_x G \supseteqq G \supseteqq [G, G] = [\widetilde{G}, \widetilde{G}]$.

2) We observe that $\widetilde{G}_x \cdot G$ is the same for all x in O. In fact, assume that a is some element in $\widetilde{G}$. Then we have:

$$\widetilde{G}_{ax} = a\widetilde{G}_x a^{-1} \subseteqq \widetilde{G}_x \cdot [G, G] \subseteqq \widetilde{G}_x G.$$

Thus $\widetilde{G}_{ax} \cdot G$ is in $\widetilde{G}_x \cdot G$ and therefore, by symmetry, $\widetilde{G}_{ax} \cdot G = \widetilde{G}_x \cdot G$. We conclude from this that if D is the closure of $\widetilde{G}_x \cdot G$ in $\widetilde{G}$, then D depends on O only.

d) Assume that O is a $\widetilde{G}$-orbit in $\mathfrak{g}^*$ such that $H \subseteqq O$ (cf. b) above). Since O is open in its closure, if $x \in O$ is fixed, then $a\widetilde{G}_x \mapsto ax$ is a homeomorphism $\widetilde{G}/\widetilde{G}_x \to O$. We conclude from this that the G-orbit closures and D orbits in O coincide. If x is in H, then $H \subseteqq \overline{Gx} \cap O$. We shall show that $H = \overline{Gx} \cap O$. In fact, if y is in $\overline{Gx} \cap O - H$, then $Dx = Dy$, $\overline{Gx} \cap O = \overline{Gy} \cap O$ and $\overline{Gx} = \overline{Gy}$. Thus $x \sim y$ and therefore $y \in H$, implying a contradiction. In this manner $H = \overline{Gx} \cap O$, and thus H is locally closed in $\mathfrak{g}^*$.

Summing up, $\sim$ is the equivalence relation, the existence of which was claimed in our Proposition.

IMPORTANT OBSERVATION. Let O be a $\widetilde{G}$-orbit, x some element of O. Take the closure of $G \cdot \widetilde{G}_x$ in $\widetilde{G}$, and let this be D. We set $G_1 = D_0$. Then for any y in O, $G_1 y$ is the R-orbit containing y. Such a G_1 will be called a symmetry group of O. Also, given x in O, the R-orbit containing x is equal to $\overline{Gx} \cap O$ (= relative closure in O). We note that G_1 is the same for each point of O ($\in \mathfrak{g}^*/\widetilde{G}$).

All this being so, we have also

LEMMA. *Let O be an R-orbit of $\mathfrak{g}^*$. Then O carries a C^∞-structure, well-determined by the condition, that the map $O \to \mathfrak{g}^*$ is an imbedding.*

In fact, we recall that, given a C^∞-manifold M, and a subset A of M with the induced topology, then A carries at most one C^∞-structure, such that the inclusion map is an imbedding. Let $x \in O$ be fixed. Transfer the C^∞-structure of $G_1/(G_1)_x$ to O. This being so, it is enough to observe that the map $G_1 \to G_1 x$ is smooth.

4.9. Generalized orbits

Our next objective will be the introduction of the generalized orbits. To this end we fix an R-orbit $\mathscr{O}$, and consider the subset

$$\mathcal{L}(\mathscr{O}) = \bigcup_{g \in \mathscr{O}} \overset{A}{G_g}$$

of $\mathcal{R}$ with the natural projection onto the base $\mathscr{O}$. As we shall show presently, $\mathcal{L}(\mathscr{O})$ carries a differentiable structure, which imposes itself. Clearly $\mathcal{L}(\mathscr{O})$ is a G-invariant subset of $\mathcal{R}$. A generalized orbit is just the closure of a G-orbit. The totality of generalized orbits establishes an equivalence relation on $\mathcal{L}(\mathscr{O})$ (see later). Let us note that the restriction of the canonical projection $\mathcal{L}(\mathscr{O}) \to \mathscr{O}$ to a G-orbit is a bijection. This is not so for generalized orbits: in fact, one can construct easily an $\mathcal{L}(\mathscr{O})$ with a positive fiber-dimension, where $\mathcal{L}(\mathscr{O})$ itself is a generalized orbit. If $\mathscr{O}$ is G-transitive or, what is the same, is a locally closed coadjoint orbit, then $\mathcal{L}(\mathscr{O})$ is just a product of $\mathscr{O}$ with a G-orbit.

We recall here the motivation that underlies these constructions. Our theory (cf. 4.7) assigns to each point p of $\mathcal{R}$ a unitary equivalence class $F(p)$ of semifinite factor representations. Writing $J(p)$ for the kernel $F(p)$ we obtain a map $J : \mathcal{R} \to \mathrm{Prim}(G)$. In 4.11, this will be shown to be surjective, and, given an element in $\mathrm{Prim}(G)$, its preimage in $\mathcal{R}$ turns out to be just a generalized orbit in the sense that we are going to define.

Below $\mathscr{O}$ will be a fixed R-orbit and G_1 a corresponding symmetry group (cf. the end of 4.8).

The following two statements are trivial.

LEMMA 1. *Assume that $y_0 \in \mathscr{O}$ is given, fixed. Then there is a neighborhood V of y_0 in $\mathscr{O}$ and $s : V \to G_1$ smooth such that $s(y)y_0 = y$ $(y \in V)$.*

We note that, evidently, we can assume that $s(y_0)$ is equal to the unity.

LEMMA 2. *Assume that a_0 is a fixed element in $\overline{G}_{y_0}$. Then there is a smooth map $a : V \to G$ such that $a(y_0) = a_0$, and $a(y) \in \overline{G}_y$ $(y \in V)$.*

In fact, with the notation of Lemma 1 it is enough to define $a(y) \equiv s(y)a(s(y))^{-1}$ $(y \in V)$.

LEMMA 3. *Assume that U is open in $\mathscr{O}$ and that there is given a smooth map $f : U \to G$ such that $f(y) \in (G_y)_0$ for all $y \in U$. Then the function $y \mapsto \chi_y(f(y))$ is smooth.*

In fact, the statement being local, we can choose $s : U \to G_1$ as in Lemma 1. With notation as loc. cit. it is enough to observe that

$$\chi_y(f(y)) = \chi_{s(y)y_0}(f(y)) = \chi_{y_0}((s(y))^{-1} \cdot f(y) \cdot s(y))$$

and thus it suffices to note that $y \mapsto (s(y))^{-1} \cdot f(y)s(y)$ is smooth.

We observe that $g \mapsto \overline{G}_g \cdot L$ $(g \in \mathscr{O})$ is constant. In fact, to this end we remark that if $a \in G_1$, then $\overline{G}_{ag} = a\overline{G}_g a^{-1}$, which is, by virtue of $[G_1, G] = L$, the same as $\overline{G}_g \cdot L$. We recall (cf. 4.4, Lemma 12), that $\overline{G}_g \cdot L = K$.

We recall that K/K_0 is free abelian; let $\phi : K \to K/K_0$ be the canonical morphism. We select an integral basis $\{a_1, \dots, a_r\}$ in K/K_0.

REMARK. It will be left to the reader to verify that the subsequent considerations are independent of the particular choice of the basis $\{a_1, \dots, a_r\}$.

Assume that $y_0 \in \mathscr{O}$ is given. By Lemma 2, there is a neighborhood V of y_0, there are $\rho_j : V \to G$ smooth maps, such that $(\rho_j(y))_{1 \le j \le r} \in \overline{G}_y$ and

$$\phi(\rho_j(y)) \equiv a_j \quad \text{for } y \in V,\ 1 \le j \le r.$$

Such a system will be referred to as a basis on V.

NOTATION. Let $\tau : \mathcal{L}(\mathscr{O}) \to \mathscr{O}$ be the canonical projection.

DEFINITION. We define the map $T_V : \tau^{-1}(V) \to V \times T^r$ by

$$T_V(\chi) = (y; \chi(\rho_1(y)), \dots, \chi(\rho_r(y))) \quad \text{if } \tau(\chi) = y \in V.$$

Let $\{\sigma_j; 1 \le j \le r\}$ be another basis on V.

We define on $\tau^{-1}(V)$:

$$T_V'(\chi) = (y; \chi(\sigma_1(y)), \dots, \chi(\sigma_r(y))) \quad \text{if } \tau(\chi) = y.$$

Our next objective will be to show that $T_V' \circ T_V^{-1} : V \times T^r \to V \times T^r$ is smooth. In fact, assume

$$\{\omega_1, \dots, \omega_r\} \in T^r.$$

We observe that $T_V^{-1}(y; \omega_1, \dots, \omega_r)$ is the element χ of $\overset{A}{G}_y$ such that $\chi | (G_g)_0 = \chi_y, \chi(\rho_j(y)) \equiv \omega_j$ $(1 \le j \le r)$.

We write $\sigma_j(y) \equiv \eta_j(y)\rho_j(y)$ $(1 \le j \le r)$. Then $\eta_j : V \to G$ is smooth and such that $\eta_j(y) \in (G_y)_0$. Set

$$c_j(y) \equiv \chi_j(\eta_j(y)) \qquad (1 \le j \le r).$$

By Lemma 3, $c_j : V \to T$ is smooth and we have on V

$$(T_V' \circ T_V^{-1})(y; \omega_1, \dots, \omega_r) = (y; c_1(y)\omega_1, \dots, c_r(y)\omega_r).$$

Summing up, there is a differentiable structure on $\mathcal{L}(\mathscr{O})$ well-determined by: Given an open set $V \subset \mathscr{O}$, the map T_V defined by

$$T_V(\chi) = (y; \chi(\rho_1(y)), \dots, \chi(\rho_r(y)))$$

is a diffeomorphism.

We repeat emphasizing that this structure is independent of the particular choice of $\{a_1, \dots, a_r\}$ and of the symmetry group G_1.

NOTATION. Assume that $A \supset B$ are groups; let $\chi \in \mathrm{X}(A), \varphi \in \mathrm{X}(B)$; then $\chi\varphi$ will stand for $(\chi|B)\varphi$.

We define also $\overset{\ominus}{K} = \{\varphi; \varphi \in \mathrm{X}(K), \varphi|K_0 \equiv 1\}$ in which manner, given $\varphi \in \overset{\ominus}{K}, \chi \in \overset{A}{G}_y$ we form

$$\varphi\chi \in \overset{A}{G}_y.$$

Our next objective will be to show that the totality of G-orbit closures defines on $\mathcal{L}(\mathscr{O})$ an equivalence relation.

Before proceeding, we note that, evidently $\mathcal{L}(\mathscr{O})$ is also a G_1-space.

Let us set $\mathscr{H} = G_1 \times \overset{\ominus}{K}$.

LEMMA 4. *We claim that $\mathscr{H}$ operates smoothly on $\mathcal{L}(\mathscr{O})$.*

We leave the routine proof to the reader.

PROPOSITION. *There is an equivalence relation on $\mathcal{L}(\mathscr{O})$ the members of which are G-orbit closures.*

The proof will be very similar to that of the Proposition in the previous section 4.8. Given $p \in \mathcal{L}(\mathscr{O})$, we form the closure of $G\mathscr{H}_p$ in $\mathscr{H}$ ($= D(p)$, say) from which, by Lemma 4, we infer that $D(p)p = \overline{Gp}$. We observe next that given $p, q \in \mathcal{L}(\mathscr{O})$, we have $D(p) = D(q)$. In fact, assume that $q = ap$ for some a in G. Taking account of $[\mathscr{H}, \mathscr{H}] = [G, G]$, we have

$$G\mathscr{H}_q = G\mathscr{H}_{ap} = G \cdot a\mathscr{H}_p a^{-1} = G\mathscr{H}_p,$$

and hence

$$D(q) = \overline{G\mathscr{H}}_q = \overline{G\mathscr{H}}_p = D(p)$$

and thus $D(p) = D(q)$.

Writing D for the connected component of the common values of $D(p)$, it suffices to note that for any u in $\mathcal{L}(\mathscr{O}) : Du = \overline{Gu}$.

COROLLARY OF THE PROOF. *There is a closed subgroup D of $\mathscr{H}$ such that*

$$\{G\text{-orbit closures}\} = \{D\text{-orbits}\}.$$

DEFINITION. A generalized orbit of G is a G-orbit closure in $\mathcal{L}(\mathscr{O})$ for some choice of $\mathscr{O} \in \mathfrak{g}^*/R$.

We note also the following

LEMMA 5. *Given a generalized orbit in $\mathcal{L}(\mathscr{O})$, it carries a C^∞-structure, well-determined by: The inclusion map is an imbedding.*

4.10. Auxiliary facts

The objective of this section is to establish some auxiliary statements to be used later.

We write $\mathcal{T} = \mathcal{D}^\perp$ and form the direct product $B = \widetilde{G} \times \mathcal{T}$. It operates on $\mathfrak{g}^*$ by $g \mapsto ag + \sigma$ ($a \in \widetilde{G}, \sigma \in \mathcal{T}$). We put for $\Omega \in \mathfrak{g}^*/B$, fixed, containing $\mathscr{O}$,

$$\mathcal{L}(\Omega) = \bigcup_{g \in \Omega} \overset{A}{G_g}, \text{ disjoint union.}$$

We can build up on $\mathcal{L}(\Omega)$ the structure of a torus bundle, similarly as in the case of $\mathscr{O} \in \mathfrak{g}^*/R$, with Ω in place of $\mathscr{O}$. We have setwise

$$\mathcal{L}(\Omega) = \bigcup_{g \in \Omega} \overset{A}{G_g} \supset \bigcup_{g \in \mathscr{O}} \overset{A}{G_g} = \mathcal{L}(\mathscr{O}).$$

LEMMA 1. *Assume that $\mathscr{O}$ is an R-orbit, contained in Ω. Then $\mathcal{L}(\mathscr{O})$ is closed in $\mathcal{L}(\Omega)$.*

In fact, by what we saw above, it is enough to show that $\mathscr{O}$ is relatively closed in Ω. This will follow from the following statement: If $x \in \mathscr{O}$ is fixed, then $\widetilde{G}x$ is relatively closed in Ω. To this end it will suffice to establish that if $(a_n) \subset \widetilde{G}$ is such that $a_n x \to y$ in Ω, then y lies in $\widetilde{G}x$. Let us denote by π the canonical projection from $\mathfrak{g}^*$ onto $\mathfrak{g}^*/\mathcal{D}^\perp = \mathcal{D}^*$. By assumption, $a_n \pi(x)$ tends to $\pi(y)$, which is in $\widetilde{G}\pi(x)$, $\pi(y) = b\pi(x)$, say $(b \in \widetilde{G})$. Since $\widetilde{G}\pi(x)$ is locally closed, there is $(b_n), (c_n) \subset \widetilde{G}$ such that $a_n = b_n c_n, b_n \to b$ and $(c_n) \subset \widetilde{G}_{\pi(x)}$. In this manner, it is enough to show that $\widetilde{G}_{\pi(x)} x$ is closed. To see this we remark first that

$$\widetilde{G}_{\pi(x)} x = \widetilde{G}x \cap (x + \mathcal{D}^\perp)$$

is locally closed. If $m \in \widetilde{G}_{\pi(x)}$, we put $mx = x + \gamma(m)$, where γ is a continuous morphism from $\widetilde{G}_{\pi(x)}$ into $\mathcal{D}^\perp$. Hence $\gamma(\widetilde{G}_{\pi(x)})$ is a locally closed subgroup of $\mathcal{D}^\perp$. Therefore it is also closed implying that $\widetilde{G}_{\pi(x)} x$ is closed in $\mathfrak{g}^*$.

Assume that E is a fixed $\widetilde{G}$-orbit in $\widehat{L}$. We note, incidentally, that K_x is fixed if $x \in E$; let us write K for it. Let us denote by Ω the complete inverse image of E, first in $\mathcal{D}^*$ by the Kirillov map, then through $\pi : \mathfrak{g}^* \to \mathcal{D}^*$ in $\mathfrak{g}^*$. Evidently, Ω is a B-orbit in $\mathfrak{g}^*$. Let us write $F = \{\rho; \rho \in \widehat{K}$ such that $\rho | L \in E\} \subset \widehat{K}$. We recall that, by Lemma 5 in 3.2, F as a subspace of $\widehat{K}$ is locally compact, separable and Hausdorff. Let us recall also that if $p = (g, \chi) \in \mathcal{L}(\Omega)$, then $\lambda(p)$ is the unitary equivalence class of $\operatorname{ind}(\mathfrak{h}, p; K) \in F$.

LEMMA 2. *The restriction of* λ *to* $\mathcal{L}(\Omega)$ *(to be denoted again by* λ*) is continuous, open and onto*: $\mathcal{L}(\Omega) \to F$.

We start by recalling the following notational convention: if $A \supset B$ are groups and $\varphi \in \mathrm{X}(A)$ and $\psi \in \mathrm{X}(B)$ then $\varphi\psi$ stands for $(\varphi | B)\psi$.

a) We have the following actions of $\widetilde{G}, \mathcal{D}^\perp$, and $\overset{\ominus}{K}$ resp. on $\mathcal{R}$:

1) For $a \in \widetilde{G} : a(g, \chi) = (ag, a\chi)$ where $a\chi$ is obtained as in the special case $a \in G$.

2) If $\sigma \in \mathcal{D}^\perp$ we define $\psi_\sigma \in \mathrm{X}(G)$ by $d\psi_\sigma = -2\pi i \sigma$ and $\sigma(g, \chi) = (g + \sigma, \psi_\sigma \chi)$.

3) For $\varphi \in \overset{\ominus}{K} : \varphi(g, \chi) = (g, \varphi\chi)$.

The actions of the different groups pairwise commute. Writing $N = B \times \overset{\ominus}{K}$, they give rise to a representation of N by diffeomorphisms of $\mathcal{L}(\Omega)$.

b) We observe next that N acts transitively in $\mathcal{L}(\Omega)$. In fact, assume that (g, χ) and (g', χ') are given. Then there is (a, σ) in B such that $ag + \sigma = g'$, yielding $(a, \sigma)(g, \chi) = (g', \chi_1)$. Hence it is enough to note that there is φ in $\overset{\ominus}{K}$ such that $\varphi\chi_1 = \chi'$.

From here we conclude that for $p_0 \in \mathcal{L}(\Omega)$ fixed, $aN_{p_0} \mapsto ap_0$ is a diffeomorphism $N/N_{p_0} \to \mathcal{L}(\Omega)$.

c) We define an action of N on F as follows:

1) For $a \in \widetilde{G}$ we have

$$a \cdot \operatorname{ind}(\mathfrak{h}, p; K) = \operatorname{ind}(a\mathfrak{h}, ap; K).$$

2) We note that for $\sigma \in \mathcal{D}^\perp$ we have $\psi_\sigma \cdot \operatorname{ind}(\mathfrak{h}, g, \chi; K) = \operatorname{ind}(\mathfrak{h}, g + \sigma, \psi_\sigma \chi; K)$.

In fact, we recall that $\operatorname{ind}(\mathfrak{h}, g, \chi; K)$ is the holomorphic part of $\operatorname{ind}_{\overline{G}_g D\uparrow G}\omega$ where $\omega|\overline{G}_g = \chi$ and $\omega|D = \chi_f$ and we have

$$\psi_\sigma\omega|\overline{G}_g = \psi_\sigma\chi, \qquad \psi_\sigma\omega|D = \chi_f.$$

3) $\varphi \cdot \operatorname{ind}(\mathfrak{h}, g, \chi; K) = \operatorname{ind}(\mathfrak{h}, g, \varphi\chi; K)$.

We note that with respect to these actions $\lambda : \mathcal{L}(\Omega) \to F$ is equivariant. Let us note also that N operates on F transitively. In fact, let us suppose that ρ_1, ρ_2 are given in F. Then there is $a \in \widetilde{G}$ such that $a\rho_1|L = \rho_2|L$. Hence it is enough to note that there is a $\sigma \in \mathcal{D}^\perp$ leading to $\psi_\sigma a\rho_1 = \rho_2$. We conclude from this that, given g_0 in F, the map $aN_{g_0} \mapsto ag_0$ is an N-equivariant homeomorphism.

All this being so, we complete the proof of Lemma 2 as follows. What we have to prove is that $\lambda : \mathcal{L}(\Omega) \to F$ is continuous, onto, and open. Unexplained notation will be as above. Suppose that p_0 is fixed in $\mathcal{L}(\Omega)$ and let us set $q_0 = \lambda(p_0)$.

Since λ is N-equivariant, we have $N_{q_0} \supset N_{p_0}$. Assume that $a \in N$, and consider the following series of maps:

$$ap_0 \mapsto aN_{p_0} \mapsto aN_{q_0} \mapsto a\lambda(p_0) = \lambda(ap_0).$$

The first and the third are, by b) and c) above, homeomorphisms, and evidently the second is continuous and open.

4.11. Proof that J is surjective

Let us recall that, for $p \in \mathcal{R}$, we define $J(p)$ as the kernel of $F(p)$. The objective of this section will be to establish that the map $p \mapsto J(p)$ from $\mathcal{R}$ into $\operatorname{Prim}(G)$ is surjective and we have $J(p_1) = J(p_2)$ if and only if p_1 and p_2 lie in the same generalized orbit. To express the last relation, we shall write $p_1 \equiv p_2(S)$.

LEMMA 1. *Unexplained notation as above.*
Assume that Ω is a B-orbit in $\mathfrak{g}^*$. Let p be some element of $\mathcal{L}(\Omega)$. Then

$$\overline{\lambda(Gp)} = \lambda(\overline{Gp}).$$

In fact, it is enough to show that $\lambda(\overline{Gp})$ is closed in F. To this end it is enough to recall (cf. 4.7, end) that if $g \in \Omega$ is fixed, $f = g|\mathcal{D}$ and $M = G_g(G_f)_0 \cdot L$, then for any s in F, its inverse image through λ is an M-orbit. Thus, since $\lambda \cdot \mathcal{L}(\Omega) \to F$ is open, $\lambda(\overline{Gp})$ is closed. Summing up, we have proved that $\lambda(\overline{Gp}) = \overline{\lambda(Gp)}$.

LEMMA 2. *Assume that p_1, p_2 are elements in $\mathcal{L}(\Omega)$. Then $p_1 \equiv p_2(S)$ if and only if $\overline{Gp_1} = \overline{Gp_2}$ in $\mathcal{L}(\Omega)$.*

In fact, we have $p_1 \equiv p_2(S)$ if and only if there is an $\mathcal{O} \in \mathfrak{g}^*/R$ such that $\overline{Gp_1} = \overline{Gp_2}$ in $\mathcal{L}(\mathcal{O})$. Since $\mathcal{L}(\mathcal{O})$ is closed in $\mathcal{L}(\Omega)$, this is equivalent to $\overline{Gp_1} = \overline{Gp_2}$ in $\mathcal{L}(\Omega)$.

LEMMA 3. *If $p_1 \equiv p_2(S)$, then $J(p_1) = J(p_2)$.*

In fact, by what we have just seen, there is $\Omega \in \mathfrak{g}^*/B$ such that $p_1, p_2 \in \mathcal{L}(\Omega)$ and $\overline{Gp_1} = \overline{Gp_2}$. This implies that $\lambda(\overline{Gp_1}) = \lambda(\overline{Gp_2})$ and hence, by Lemma 1, $\overline{G\lambda(p_1)} = \overline{G\lambda(p_2)}$ in F. Let us put $\rho_k = \lambda(p_k)$ $(k = 1, 2)$. Then $\overline{G\rho_1} = \overline{G\rho_2}$ whence, by Lemma 10 in 3.2, $J(\rho_1) = J(\rho_2)$. All this being so, it is enough to note that $J(p) = \ker(\operatorname{ind}_{K\uparrow G} \lambda(p))$.

LEMMA 4. *Assume that p_1 and p_2 are such that $J(p_1) = J(p_2)$. We claim that $p_1 \equiv p_2(S)$.*

In fact, let us suppose that $p \in \mathcal{R}$, $p = (g, \chi)$. If $T \in F(p)$, then there is $U \in \lambda(p)$ such that $T = \mathrm{ind}_{K \uparrow G} U$.

We have that $U|L$ corresponds in the sense of Kirillov, to $L(g|\mathcal{D})$. Let us recall that there is $E(J) \in \widehat{L}/\widetilde{G}$ such that if $T \in \mathrm{Fac}(G)$ is such that $\ker(T) = J$, then $T|L$ is carried by $E(J)$. Let us put $J_k = J$ $(k = 1, 2)$. Let, as above, Ω be the inverse image of $E(J)$ in $\mathfrak{g}^*$.

Assume that $p_k = (g_k, \chi_k)$ $(k = 1, 2)$; thus $g_k \in \Omega$ and $p_k \in \mathcal{L}(\Omega)$. We set $\rho_k = \lambda(p_k)$. We have by assumption $J(\rho_1) = J(\rho_2)$ whence, by Lemma 9 in 3.2, $\overline{G\rho_1} = \overline{G\rho_2} \Rightarrow$ (by Lemma 1) $\overline{Gp_1} = \overline{Gp_2}$ implying $p_1 \equiv p_2(S)$.

LEMMA 5. *We claim that* $J : \mathcal{R} \to \mathrm{Prim}(G)$ *is surjective.*

This is implied by the Proposition in 3.3.

Summing up, we have proved

PROPOSITION. *Suppose that G is connected, simply connected, and solvable. Then the map $J : \mathcal{R} \to \mathrm{Prim}(G)$ is surjective. We have $J(p_1) = J(p_2)$ if and only if p_1, p_2 lie on the same generalized orbit.*

4.12. Construction of a normal representation with kernel J

We recall (cf. 3.8, Theorem) that given a connected Lie group G, an element J of $\mathrm{Prim}(G)$, then there is a normal representation T of G, the kernel of which is equal to J. Furthermore T, up to quasi-equivalence, is well-determined by J. Next we shall assume that G is connected, simply connected, and solvable. The objective of this section is to construct T by aid of the orbit picture.

Assume that O is a fixed generalized orbit. We claim that there is $\{T(p); p \in O\}$ measurable, such that $T(p) \in F(p)$ for all p. In fact, it is enough to show this locally.

(i) We note first that, given $T(p) \in F(p)$ and $\varphi \in \overset{\ominus}{K}$, we can define $\varphi T(p)$ as follows.

Assume that $\tilde{\varphi} \in \mathrm{X}(G)$ is such that $\tilde{\varphi}|K \equiv \varphi$. We recall that $T(p) = \mathrm{ind}_{K \uparrow G} \lambda(p)$. This being so we can define

$$\varphi T(p) \underset{\mathrm{def}}{=} \tilde{\varphi} T(p) = \mathrm{ind}_{K \uparrow G} \varphi \lambda(p).$$

(ii) Let the Lie group D, corresponding to O be as in 4.9. Assume that p_0 is a fixed point of O. We select a neighborhood V of p_0 for which there is a smooth map $d : V \to D$ satisfying $d(p) \cdot p_0 \equiv p$ $(p \in V)$. Let us assume that $d(p) \equiv (a(p), \varphi(p)) \in G_1 \times \overset{\ominus}{K}$. If T_0 is fixed in $F(p_0)$ and we set $T(p) = \varphi(p) a(p) T_0$, then $T(p) \in F(p)$ $(p \in V)$ and $p \mapsto T(p)$ is measurable.

Let v be an absolutely continuous measure on O. We form

$$T = \int_O^{\oplus} T(p) \cdot dv(p).$$

By 2.4 Lemma and 4.7, this is a factor representation. Let $\mathcal{F}(O)$ be its unitary equivalence class. By 4.11, we can define unambiguously $J(O) \equiv J(p)$ $(p \in O)$, and $O \mapsto J(O)$ is a bijection between the totality of generalized orbits and the underlying set of $\mathrm{Prim}(G)$. Clearly we have $\ker(\mathcal{F}(p)) = J(O)$. Our objective below will be to show that if $T \in \mathcal{F}(O)$, then T is a normal representation. By doing so, we shall have accomplished the goal set forth at the start of this section.

Let $\Omega \in \mathfrak{g}^*/B$ be such that $O \subset \mathcal{L}(\Omega)$ (cf. 4.10) and $\mathfrak{t}$ the Σ-orbit in F such that $\lambda(O) = \mathfrak{t}$ (cf. 3.2, end). We recall that if $\mathfrak{t} \to \operatorname{Irr}(K)$ is a section, μ absolutely continuous, and we put

$$T = \operatorname{ind}_{K\uparrow G}\left(\int_{\mathfrak{t}}^{\oplus} \rho(\zeta)\cdot d\mu(\zeta)\right)$$

then T is a normal representation such that $\ker(T) \equiv J(\rho)$ $(\rho \in \mathfrak{t})$, where $J(\rho) = \ker(\operatorname{ind}_{K\uparrow G}\rho)$. We put $T(p) = \operatorname{ind}_{K\uparrow G}\rho(\lambda(p))$. We have $T(p) \in F(p)$.

We set

$$T' = \int_O^{\oplus} T(p)\cdot dv(p).$$

Then $T' \in \mathcal{F}(O)$, and it will suffice to show that T and T' are quasi-equivalent. Evidently, we can assume that $\mu = \lambda(v)$ and form

$$v = \int_{\mathfrak{t}} v_\zeta \cdot d\mu(\zeta).$$

Here v_ζ is carried by $\lambda^{-1}(\{\zeta\})$. We have

$$T' = \int_O^{\oplus} T(p)\cdot dv(p) = \operatorname{ind}_{K\uparrow G}\left(\int_O^{\oplus} \rho(\lambda(p))\cdot dv(p)\right).$$

But

$$\int_O^{\oplus} \rho(\lambda(p))\cdot dv(p) = \int_{\mathfrak{t}}\left(\int_O \rho(\lambda(p))\cdot dv_\zeta(p)\right) d\mu(\zeta).$$

Since we have on $\lambda^{-1}(\{\zeta\}) : \rho(\lambda(p)) \equiv \rho(\zeta)$, $\int_O^{\oplus} \rho(\lambda(p))\cdot dv_\zeta(p)$ is quasi-equivalent to $\rho(\zeta)$ and hence so is $\int_O^{\oplus} \rho(\lambda(p))\cdot dv(p)$ with $\int_{\mathfrak{t}}^{\oplus} \rho(\zeta)\cdot d\mu(\zeta)$.

We can conclude from all this that $T' = \int_O^{\oplus} T(p)\cdot dv(p)$ is quasi-equivalent to

$$\operatorname{ind}_{K\uparrow G}\left(\int_{\mathfrak{t}}^{\oplus} \rho(\zeta)\cdot d\mu(\zeta)\right) = T.$$

Summing up we have the following

THEOREM. *Suppose that J is a given primitive ideal of $C^*(G)$. Assume that O is the generalized orbit such that $J(p) \equiv J$ ($p \in O$; cf. 4.11). Assume that $T : p \mapsto T(p) \in F(p)$ is measurable and v absolutely continuous on O. Form*

$$T = \int_O^{\oplus} T(p)\cdot dv(p).$$

Then T is a normal representation with a kernel equal to J.

4.13. Type-one primitive ideals

We recall (cf. 3.6), that given $J \in \operatorname{Prim}(G)$ we say that J is of type one (and write $J = (I)$) if and only if one of the two following conditions is met.

1) If T is a factor representation of G such that $\ker(T) = J$, then T is of type I.

2) There is an irreducible normal representation belonging to J.

In this concluding section of our chapter, we will characterize type-one primitive ideals.

To this end we recall first that if g is some element of $\mathfrak{g}^*$, π an irreducible representation of L belonging, in the sense of Kirillov, to $L(g|D)$, $K = K_\pi$ $(= L\overline{G}_g$;

cf. Lemma 12 in 4.4), $\rho \in \widehat{K}$ such that $\rho|L = \pi$, then by Lemma 8 in 3.1 and Lemma 13 in 4.4, $\mathrm{ind}_{K\uparrow G}\,\rho$ is of type I if and only if $G_g/\overline{G}_g$ is finite.

All this being so, we have the

PROPOSITION. *Assume that J is a primitive ideal of G. Let $\mathcal{O}$ be the corresponding R-orbit. Then J is of type one if and only if*

1) *$\mathcal{O}$ is G-transitive;*
2) *for any g in $\mathcal{O}$, $G_g/\overline{G}_g$ is finite.*

In fact, let us assume first that the two conditions above are fulfilled and prove that J is of type I.

Let O be a generalized orbit over $\mathcal{O}$. Then O is G-transitive. Hence, if $p \in O$, by 4.12, $T(p)$ is normal. But, by $T(p) = \mathrm{ind}_{K\uparrow G}\,\lambda(p)$, $T(p)$ is of type one if and only if $G_g/\overline{G}_g$ is finite.

To conclude the proof of our Proposition, let us assume now that the primitive ideal J is of type one and show that in this case $\mathcal{O}$ is G-transitive and $G_g/\overline{G}_g$ is finite. In fact, let O be the generalized orbit corresponding to J. We claim that O is G-transitive. In fact, we start by repeating that if J is of type I, then there is an irreducible, normal representation ρ such that $\ker(\rho) = J$. We recall also that $O = \{p; p \in \mathcal{R}$ such that $J(p) = J\}$. If $p \in O$ and $T(p) \in F(p)$, then (by $\ker(T(p)) = J(p) = J$) $T(p)$ is of type one and hence (cf. [**D***], p. 109) $T(p)$ is a multiple of an irreducible representation σ of G. By $\ker(\rho) = J = \ker(\sigma)$ (cf. [**D***], p. 85), ρ and σ are unitarily equivalent. In this manner, if $p, q \in O$, then $F(p) \approx F(q)$ and thus by 4.7, $Gp = Gq$.

We have thus shown that if J is of type one, then the corresponding R-orbit is G-transitive. By what we said above, evidently, in this case also $G_g/\overline{G}_g$ is finite. We derive from all this the

THEOREM (Auslander-Kostant). *Let G be connected, simply connected, solvable. Then G is of type one if and only if any coadjoint orbit is locally closed, and, for any $g \in \mathfrak{g}^*$, χ_g extends to G_g.*

In fact, this is a trivial implication of the Proposition above jointly with the observation that G is of type one if and only if any element of $\mathrm{Prim}(G)$ is of type one.

References

[AK] L. Auslander and B. Kostant, *Polarization and unitary representations of solvable Lie groups*, Invent. Math. **14** (1971), 255–354.

[AM] L. Auslander and C. C. Moore, *Unitary representations of solvable Lie groups*, Memoirs of the Amer. Math. Soc., no. 62, 1966.

[B] N. Bourbaki, *Algèbres de Lie*, Act. Sci. Ind. no. 1285, Hermann, Paris, 1960.

[B int] N. Bourbaki, *Intégration vectorielle*, Act. Sci. Ind. no. 1281, Hermann, Paris, 1959.

[Bl] R. J. Blattner, *Positive definite measures*, Proc. Amer. Math. Soc. **14** (1968), 423-428.

[Ch II] C. Chevalley, *Théorie des groupes de Lie*, vol. 2, Act. Sci. Ind., no. 1152, Hermann, Paris, 1951.

[Ch III] C. Chevalley, *Théorie des groupes de Lie*, vol. 3, Act. Sci. Ind., no. 1226, Hermann, Paris, 1955.

[C] A. Connes, *Classification of injective factors*, Ann. of Math. **104** (1976), 73–115.

[D] J. Dieudonné, *Eléments d'Analyse*, Gauthier-Villars, Paris, 1974.

[Dq] J. Dixmier, *Algèbres quasi-unitaires*, Comment. Math. Helv. **26** (1952), 275–322.

[D vN] J. Dixmier, *Les algèbres d'opérateurs dans l'espace Hilbertien* (*Algèbres de von Neumann*), 2nd edition, Gauthier-Villars, Paris, 1969.

[Da] J. Dixmier, *Sur les représentations unitaires des groupes de Lie algébriques*, Ann. Inst. Fourier **7** (1957), 315–328.

[D*] J. Dixmier, *Les C^*-algèbres et leurs représentations*, Gauthier-Villars, Paris, 1964.

[Dh] J. Dixmier, *Représentations induites holomorphes des groupes résolubles algébriques*, Bull. Soc. Math. France **94** (1966), 181–206.

[Dal] J. Dixmier, *Sur les C^*-algèbres*, Bull. Soc. Math. France **88** (1960), 95–112.

[Dreg] J. Dixmier, *Sur la représentation régulière d'un groupe localement compact connexe*, Ann. Sci. École Norm. Sup. **2** (1969), 423–436.

[E] E. G. Effros, *A decomposition theory for representations of C^*-algebras*, Trans. Amer. Math. Soc. **107** (1963), 83–106.

[EH] E. G. Effros and F. Hahn, *Locally compact transformation groups and C^*-algebras*, Memoirs of the Amer. Math. Soc., no. 75 (1967).

[F1] J. M. G. Fell, *Weak containment and induced representations of groups*, Canad. J. of Math. **14** (1962), 237–268.

[F2] J. M. G. Fell, *Weak containment and Kronecker products of group representations*, Pacific Journ. of Math. **13** (1963), 503–510.

[Gl] J. Glimm, *Locally compact transformation groups*, Trans. Amer. Math. Soc. **101** (1961), 124–138.

[Go] R. Godement, *Mémoire sur la théorie des caractères dans les groupes localement compacts unimodulaires*, Journ. de Math. Pures et Appl. **30** (1951), 1–110.

[Grl] F. P. Greenleaf, *Invariant means of locally compact groups and their applications*, van Nostrand, New York, 1969.

[H] Harish-Chandra, *Representations of semi-simple Lie groups*, Trans. Amer. Math. Soc. **76** (1954), 234–253.

[Hal] H. Halpern, *Quasi-equivalence classes of normal representations for a separable C^*-algebra*, Trans. Amer. Math. Soc. **203** (1975), 129–140.

[Hochs] G. Hochschild, *The structure of Lie groups*, Holden-Day Ser. Math., 1965.

[Kir I] A. A. Kirillov, *Unitary representations of nilpotent Lie groups*, Usp. Math. Nauk **17** (1962), 57–110.

[Kir II] A. A. Kirillov, *Characters of unitary representations of Lie groups, Reduction theorems*, Functional Analysis and its Applications **3** (1969), 36–47.

[Mackann I] G. W. Mackey, *Induced representations of locally compact groups.* I, Ann. of Math. **55** (1952), 101–139.

[Mackann II] G. W. Mackey, *Induced representations of locally compact groups.* II, Ann. of Math. **58** (1953), 193–222.

[Macktr] G. W. Mackey, *Borel structure in groups and their duals*, Trans. Amer. Math. Soc. **85** (1957), 134–165.

[Mackact] G. W. Mackey, *Unitary representation of group extensions*, Acta Math. **99** (1958), 265–311.

[Mack Jer] G. W. Mackey, *Induced representations of normal subgroups*, Proc. of the Internat. Symposium on Linear Spaces, Jerusalem, 1966, 319–326.

[N] J. von Neumann, *Die Eindeutigkeit der Schrödingenschen Operatoren*, Math. Ann. **104** (1931), 570–578.

[Pog] D. Poguntke, *Simple quotients of group C^*-algebras for two step nilpotent groups and connected Lie groups*, Ann. Sci. École Norm. Sup. **16** (1983), 151–172.

[Lp 1] L. Pukánszky, *Leçons sur la représentation des groupes*, Dunod, Paris, 1966.

[Lp 2] L. Pukánszky, *Unitary representation of solvable Lie groups*, Ann. Sci. École Norm. Sup. **4** (1971), 457–608.

[Lp 3] L. Pukánszky, *Action of algebraic groups of automorphisms on the dual of a class of type* I *groups*, Ann. Sci. École Norm. Sup. **5** (1972), 379–396.

[Lp 4] L. Pukánszky, *The primitive ideal space of solvable Lie groups*, Invent. Math. **22** (1973), 74–118.

[Lp 5] L. Pukánszky, *Characters of connected Lie groups*, Acta Math. **133** (1973), 81–137.

[Lp 6] L. Pukánszky, *Symplectic structure in generalized orbits of solvable Lie groups*, Journ. Reine Angew. Math. **347** (1984), 33–68.

[Lp 7] L. Pukánszky, *Quantization and Hamiltonian G-foliations*, Trans. Amer. Math. Soc. **295** (1986), 811–847.

[S] I. E. Segal, *An extension of Plancherel's formula to separable unimodular groups*, Ann. of Math. **52** (1950), 272–297.

[W] A. Weil, *L'intégration dans les groupes topologiques et ses applications*, Hermann, Paris, 1953.

Selected Titles in This Series

(Continued from the front of this publication)

40.4 **Daniel Gorenstein, Richard Lyons, and Ronald Solomon,** The classification of the finite simple groups, number 4, 1999

40.3 **Daniel Gorenstein, Richard Lyons, and Ronald Solomon,** The classification of the finite simple groups, number 3, 1998

40.2 **Daniel Gorenstein, Richard Lyons, and Ronald Solomon,** The classification of the finite simple groups, number 2, 1995

40.1 **Daniel Gorenstein, Richard Lyons, and Ronald Solomon,** The classification of the finite simple groups, number 1, 1994

39 **Sigurdur Helgason,** Geometric analysis on symmetric spaces, 1994

38 **Guy David and Stephen Semmes,** Analysis of and on uniformly rectifiable sets, 1993

37 **Leonard Lewin, Editor,** Structural properties of polylogarithms, 1991

36 **John B. Conway,** The theory of subnormal operators, 1991

35 **Shreeram S. Abhyankar,** Algebraic geometry for scientists and engineers, 1990

34 **Victor Isakov,** Inverse source problems, 1990

33 **Vladimir G. Berkovich,** Spectral theory and analytic geometry over non-Archimedean fields, 1990

32 **Howard Jacobowitz,** An introduction to CR structures, 1990

31 **Paul J. Sally, Jr. and David A. Vogan, Jr., Editors,** Representation theory and harmonic analysis on semisimple Lie groups, 1989

30 **Thomas W. Cusick and Mary E. Flahive,** The Markoff and Lagrange spectra, 1989

29 **Alan L. T. Paterson,** Amenability, 1988

28 **Richard Beals, Percy Deift, and Carlos Tomei,** Direct and inverse scattering on the line, 1988

27 **Nathan J. Fine,** Basic hypergeometric series and applications, 1988

26 **Hari Bercovici,** Operator theory and arithmetic in H^∞, 1988

25 **Jack K. Hale,** Asymptotic behavior of dissipative systems, 1988

24 **Lance W. Small, Editor,** Noetherian rings and their applications, 1987

23 **E. H. Rothe,** Introduction to various aspects of degree theory in Banach spaces, 1986

22 **Michael E. Taylor,** Noncommutative harmonic analysis, 1986

21 **Albert Baernstein, David Drasin, Peter Duren, and Albert Marden, Editors,** The Bieberbach conjecture: Proceedings of the symposium on the occasion of the proof, 1986

20 **Kenneth R. Goodearl,** Partially ordered abelian groups with interpolation, 1986

19 **Gregory V. Chudnovsky,** Contributions to the theory of transcendental numbers, 1984

18 **Frank B. Knight,** Essentials of Brownian motion and diffusion, 1981

17 **Le Baron O. Ferguson,** Approximation by polynomials with integral coefficients, 1980

16 **O. Timothy O'Meara,** Symplectic groups, 1978

15 **J. Diestel and J. J. Uhl, Jr.,** Vector measures, 1977

14 **V. Guillemin and S. Sternberg,** Geometric asymptotics, 1977

13 **C. Pearcy, Editor,** Topics in operator theory, 1974

12 **J. R. Isbell,** Uniform spaces, 1964

11 **J. Cronin,** Fixed points and topological degree in nonlinear analysis, 1964

10 **R. Ayoub,** An introduction to the analytic theory of numbers, 1963

9 **Arthur Sard,** Linear approximation, 1963

8 **J. Lehner,** Discontinuous groups and automorphic functions, 1964